Probability Theory for Fuzzy Quantum Spaces with Statistical Applications

Authored by

Renáta Bartková

Podravka International s.r.o, Zvolen,
Slovakia

Beloslav Riečan

Department of Mathematics, Faculty of Natural Sciences,
Matej Bel University, Tajovského 40, SK-974 01 Banská Bystrica,
Slovakia

Mathematical Institute Slovak Academy of Sciences,
Štefánikova 49, SK-814 73 Bratislava,
Slovakia

Anna Tirpáková

Department of Mathematics, Faculty of Natural Sciences,
Constantine the Philosopher University, Trieda A. Hlinku 1, SK-949 74 Nitra,
Slovakia

Probability Theory for Fuzzy Quantum Spaces with Statistical Applications

Authors: Renáta Bartková, Beloslav Riečan and Anna Tirpáková

eISBN (Online): 978-1-68108-538-8

ISBN (Print): 978-1-68108-539-5

© 2017, Bentham eBooks imprint.

Published by Bentham Science Publishers – Sharjah, UAE. All Rights Reserved.

General:

1. Any dispute or claim arising out of or in connection with this License Agreement or the Work (including non-contractual disputes or claims) will be governed by and construed in accordance with the laws of the U.A.E. as applied in the Emirate of Dubai. Each party agrees that the courts of the Emirate of Dubai shall have exclusive jurisdiction to settle any dispute or claim arising out of or in connection with this License Agreement or the Work (including non-contractual disputes or claims).
2. Your rights under this License Agreement will automatically terminate without notice and without the need for a court order if at any point you breach any terms of this License Agreement. In no event will any delay or failure by Bentham Science Publishers in enforcing your compliance with this License Agreement constitute a waiver of any of its rights.
3. You acknowledge that you have read this License Agreement, and agree to be bound by its terms and conditions. To the extent that any other terms and conditions presented on any website of Bentham Science Publishers conflict with, or are inconsistent with, the terms and conditions set out in this License Agreement, you acknowledge that the terms and conditions set out in this License Agreement shall prevail.

Bentham Science Publishers Ltd.
Executive Suite Y - 2
PO Box 7917, Saif Zone
Sharjah, U.A.E.
Email: subscriptions@benthamscience.org

**BENTHAM
SCIENCE**

CONTENTS

FOREWORD

This publication is a result of cooperation of three generations of authors: Dr. h. c. mult. prof. RNDr. Beloslav Riečan, DrSc., prof. RNDr. Anna Tirpáková, CSc. and Mgr. Renáta Bartková, PhD. I have been related with all the three of them through cooperation during several years.

From 1985 to 1989 we were colleagues with professor B. Riečan at the Department of Mathematics of Military College in Liptovský Mikuláš. That time I was just a novice lecturer, but professor B. Riečan joined our department as a distinguished and highly regarded Slovak mathematician. In that period his scientific work was focused on theory of measure and integral on ordered structures, and probability theory on non-Boolean structures - orthomodular lattices, posets, so called quantum logics. He also guided some of his new colleagues into these branches, namely B. Harman, J. Haluška, F. Chovanec, M. Jureèková, F. Kôpka, P. Malièký, J. Rybárik, E. Rybáriková-Drobná. It was in those times, in the second half of 1980s, that B. Riečan suggested introducing the notion of uncertainty based on fuzzy sets principle into quantum logics. He defined the term of F-quantum space (fuzzy quantum space) where the main terms were F-state and F-observable. F-quantum spaces became the subject of intensive study of group known as Slovak school of fuzzy sets whose standard bearers besides B. Riečan were A. Dvurečenskij and R. Mesiar. Besides the above mentioned personalities from Liptovský Mikuláš several others gradually joined the group, namely S. Bodjanová (University of Economics in Bratislava), J. Bán and M. Kalina (Faculty of Mathematics, Physics and Informatics of Comenius University in Bratislava), A. Kolesárová and M. Šabo (Slovak University of Technology in Bratislava), D. Markechová (UKF Nitra), A. Tirpáková (Institute of Archaeology, Slovak Academy of Sciences in Nitra), B. Stehlíková (District Housing Enterprise Nové Zámky), P. Vojtáš, R. Friè, M. Papèo (Mathematical Institute, Slovak Academy of Sciences in Bratislava - separate unit in Košice), M. Gavalec (Technical University of Košice), V. Janiš (Matej Bel University in Banská Bystrica), A. Marková-Stupòanová (Slovak University of Technology in Bratislava). During his stay in Liptovský Mikuláš professor Riečan chose for another meeting point of people from all around the former Czechoslovakia interested in fuzzy sets the nearby valley Jánska dolina, and thus he contributed in great extent to the emergence of Czech-Slovak and later on Czech-Slovak-Polish conferences Theory and Applications of Fuzzy Sets. After the change of political system in Czechoslovakia these trilateral conferences grew bigger and became international biennial conferences FSTA (Fuzzy Sets Theory and Applications) which have regularly taken place in Liptovský Ján since 1992. During the first FSTA conference F. Kôpka, at that time a postgraduate student supervised by B. Riečan, proposed fuzzy quantum model in which the primary operation is the partial operation of difference of fuzzy sets. He named this model a difference poset of fuzzy sets (D-poset of fuzzy sets). D-posets of fuzzy sets as well as their generalized abstract structure, difference posets (D-posets), were highly accepted by international community of mathematicians. D-posets interrelated algebraic structures which had seemed to be mutually unrelated until that time - quantum logics and multiple-valued logics (MV-algebras). Unfortunately, in 2008 a severe illness untimely ended life of F. Kôpka. In order to honour his memory, B. Riečan named D-poset of fuzzy sets with another special operation of product of fuzzy sets as Kôpka's D-poset. In further period professor Riečan together with his postgraduate students dedicated themselves mostly to development of probability theory on MV-algebras and IF-sets. In September 1987 I together with my colleague F. Kôpka attended an internship at the Mathematical Institute of Slovak Academy of Sciences in Bratislava under the supervision of an expert professor A. Dvurečenskij. There we met A. Tirpáková, a postgraduate student supervised by professor Dvurečenskij. Together we studied quantum structures, fuzzy sets theory, and explored F-quantum spaces introduced by B. Riečan. A. Tirpáková was mainly focused on F-observables, their summability, various types of their convergences; she studied ergodic theory on F-quantum spaces and achieved remarkable results in this branch. Since then we regularly met on scientific conferences, such as PROBASTAT, Winter School on Measure Theory, Theory and Applications of Fuzzy Sets, FSTA and Nitra Statistical Days.

The third author, R. Bartková, started her postgraduate study under supervision of professor Riečan in 2011, but then the supervision of her study was passed to me. In her dissertation she devoted herself to study of validity of fundamental theorems on extreme values on various algebraic structures: non-additive probability space, fuzzy quantum space, MV-algebras and IF-events. In addition to the above mentioned theoretical contribution, she also achieved practical contribution to statistical processing of concrete data from IF-sets by means of principal component analysis and factor analysis in order to reduce the dimension of data set. The results she achieved are included in several chapters of this publication.

The authors of publication Probability Theory for Fuzzy Quantum Spaces with Statistical Applications introduce the reader into the issues of probability theory, gradually from the traditional Kolmogorov probability space where the domain is *set al*gebra, then Zadeh space focused on fuzzy sets, continuing with probability theory on IF-sets, MV-algebras, and finally on F-quantum spaces. The reader can get a coherent picture about the issues of probability theory on particular algebraic structures, and thus become able to judge mutual relationships and differences.

Thanks to high scientific and methodical erudition of the authors the publication is written in an accessible and comprehensible style. I am confident that it is going to be a sought-after study material for students, postgraduate students, and for all those interested in the given issues.

doc. RNDr. Ferdinand Chovanec, CSc.
Department of Natural Sciences
Armed Forces Academy
Liptovký Mikuláš, Slovakia

PREFACE

Probability theory, like other mathematical disciplines, underwent a tumultuous development in the past. In 1933 Kolmogorov published work [39], in which he introduced the axiomatic model of his probability theory. In classical probability theory based on Kolmogorov axiomatic model it is assumed that the events associated with the experiment form the Boolean σ-algebra of subsets S of the set Ω. Probability is then the σ-additive nonnegative final function P on S with values in interval $[0,1]$, whereby it holds that if $\{A_n\}$ is a sequence of mutually exclusive events from S, then $P(\bigcup_n A_n) = \sum_n P(A_n)$ and $P(\Omega) = 1$.

At present, we can say that Kolmogorov axioms deeply infiltrated not only into probability theory and mathematical statistics, but they also encouraged the development of other scientific disciplines, such as physics, biology, economics, social sciences, *etc*.

Over time, however, it has emerget that for some scientific disciplines, for example quantum mechanics, the concept of σ-algebra is too limiting. It does not describe such situations that arise in connection with Heisenberg uncertainty principle, which claims that if any two observables (such as a position x and a momentum y) are measured at the same time, the product of the squares of errors of measurement Δx and Δy is connected with the inequality $(\Delta x)^2 \cdot (\Delta y)^2 \geq h^2 > 0$, *i.e.* the accuracy of measurement of one variable happens at the expense of the second one. Birkhoff and von Neumann [5] pointed to the fact that the set of experimentally verifiable statements about quantum-mechanical system has different algebraic structure than the Boolean algebra in [39]. The first attempts at the mathematical formulation of the quantum mechanics come from Heisenberg [35] and Schrödinger [96]. Heisenberg proposed the formalism of matrix mechanics and Schrödinger the one of wave mechanics. Both theories were generalized by von Neumann [63]. He proposed a model of quantum mechanics based on a complex separable Hilbert space.

Nowadays, one of the most used axiomatic models of quantum mechanics is quantum logic, which recorded a rapid development in 1960, when works of Varadarajan [103], Mackey [43], Mac Laren [42], Gunson [29] and others appeared.

The basic mathematical model of the current quantum theory is the von Neumann model, based on the geometry of Hilbert space (Varadarajan, [103]). If we define the system M of all closed subspaces of the given Hilbert space (where the notion "a state of system" means a measure of probability on $\mathcal{M}$ according to Varadarajan [103]), and this definition of the state is compared with the definition of P-measure on fuzzy sets (according to Piasecki [70]), we can see that both objects have similar algebraic structure. In 1985 Piasecki [70] submitted a model called soft σ-algebra in the fuzzy set theory. This model demonstrated several identical characteristics of this new structure with quantum logics. That analogy, which was for the first time observed by Riečan [79] and later by Pykacz [73], led us to the idea to build a quantum theory based on fuzzy sets.

The theory of fuzzy sets originated in 1960s in connection with the emergence of article by Zadeh [106]. If we return to the analogy between quantum logics and fuzzy sets theory, according to Dvurečenskij [13] a fuzzy set can be viewed as a fuzzy event, respectively as a real-valued function, which is defined on set $X (= \Omega)$ with values in the interval [0,1], which describes the fuzziness of set (event) a within the meaning of Zadeh [106]. Number $a(x)$ indicate the measure of membership of point x to set a. If X is a non-empty set, called the universum, and $\mathcal{M}$ is a system of fuzzy subsets of universum X, *i.e.* the system of functions on X with values in interval $[0, 1]$, then according to Riečan [79] we say that $(X, \mathcal{M})$ is an F- quantum space, according to Dvurečenskij and Chovanec [17] also called a fuzzy quantum space, or according to Dvurečenskij [12] a fuzzy measurable space. The set $\mathcal{M}$ according to Piasecki [70] is also called a soft σ-algebra.

More general structures of fuzzy quantum spaces were studied by Dvurečenskij and Chovanec [16, 17] and Dvurečenskij, Chovanec and Kôpka [18]. The law of exclusions of the third or orthomodular law does not apply on set $\mathcal{M}$, which according to Mittelstaedt [57] can play certain role in axiomatic models of quantum mechanics.

In the theory of fuzzy quantum spaces many authors try to prove some known assertions from the classical probability theory. For example, the existence of fuzzy state on fuzzy quantum space was studied by Dvurečenskij [12], Navara [59] and Navara, Pták [60, 61], joint fuzzy observables and joint distributions of fuzzy observables were studied by Dvurečenskij and Riečan [21, 22]. Dvurečenskij, Kôpka and Riečan [19] proved the representation theorem, which also includes the case in fuzzy quantum space. The theory of an indefinite integral on the fuzzy quantum space was studied by Riečan [82, 83] and the extension of the validity of the Bayes formula for fuzzy sets was investigated by Mesiar [53-55], Piasecki [68, 69] Piasecki and Svitalski [71]. The entropy on fuzzy quantum space was studied by Markechová [46, 48, 49].

An important fact for the study of many assertions in the fuzzy sets theory is the existence of sum of fuzzy observables. Harman and Riečan [34] proved the existence of the sum of compatible fuzzy observables.

Among the important concepts of probability theory belong various types of convergence of random variables, which are important especially for those parts which deal with the validity of various forms of the law of large numbers and the central limit theorem. Thus the problem of generalizations of different types of convergence for fuzzy quantum space $(X, \mathcal{M})$ became topical. Several authors studied particular types of convergences on quantum logic. We will mention only those works which were the basic material for the study of various types of convergences of fuzzy observables on fuzzy quantum space $(X, \mathcal{M})$, and that is Dvurečenskij and Pulmanová [20], Jajte [37], Ochs [64, 65], Cushen [9], Gudder [26], Révesz [76]. Dvurečenskij [10], Riečan [80, 81], Chovanec and Kôpka [36], Kôpka and Chovanec [40], and others dealt with some types of convergences of fuzzy observables on fuzzy quantum space.

The issue of the ergodic theory on quantum logics was studied by more authors. Here we present only those works which were the basis for the generalization of the ergodic theory for fuzzy quantum space $(X, \mathcal{M})$: Pulmanová [72], Dvurečenskij [15] and Mesiar [56]. The first authors to prove the individual ergodic theorem for compatible fuzzy observables on fuzzy quantum space were Harman and Riečan [34]. The proof of the individual ergodic theorem for more general case is provided in this book.

The Hahn-Jordann decomposition on quantum logics exists only in partial cases, for example it exists on the logic of Hilbert space (Šerstnev [99], Dvurečenskij [14], Rüttimann [94]). The Hahn-Jordann decomposition on fuzzy quantum space is studied in this book (Chapter 4).

The following figure displays the position of fuzzy quantum space in various algebraic structures.

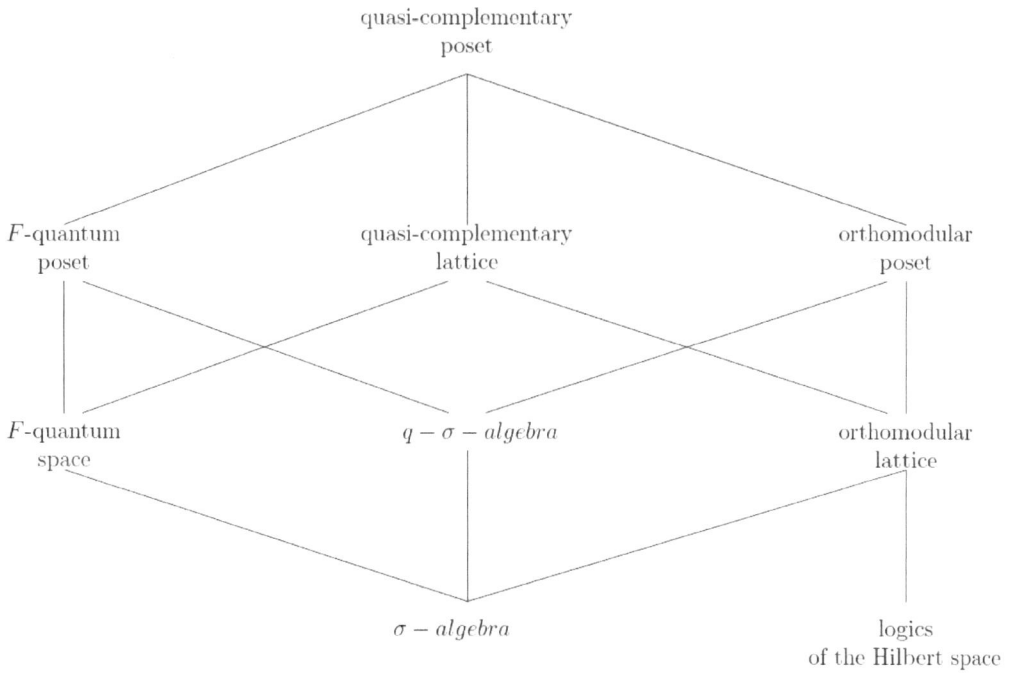

Source: [84]

Fig. An F-quantum space in various algebraic structures.

ACKNOWLEDGMENT

Hereby we would like to thank prof. Ferdinand Chovanec for his Foreword, and Tomáš Škraban, PhD. and Zuzana Naštická, M.A. for reviewing the English language correctuess. We also acknowledge the Bentham Science Publishers and referees for their remarks and recommendations.

CONFLICT OF INTEREST

The authors confirm that this eBook content has no conflict of interest.

ABOUT THE AUTHORS

Beloslav Riečan was born in Žilina, Slovakia, in November, 1936. He received his RNDr. degree in Mathematics at Comenius University in Bratislava in 1965, his PhD. degree in 1965 at the Slovak Academy of Sciences and his DrSc. degree in 1980 at Comenius University. He was Assistant Professor in 1967 and Professor in 1981 both at Comenius University. In 1958 - 1972 he worked at Slovak Technical University, in 1972 - 1985 and 1990 - 1993 at Comenius University, in 1985 - 1989 at Slovak Military Academy, and since 1998 at Matej Bel University in Banská Bystrica.

In 1990 he was the Dean of the Faculty of Mathematics and Physics at Comenius University; during 1992 - 1998 he was the Head of the Mathematical Institute of the Slovak Academy of Sciences.

His research work is concentrated on measure theory, probability theory, fuzzy sets theory and the mathematical theory of music. He published 8 monographs, around 270 scientific papers, many textbooks and many papers for a larger community.

He received many distinctions; he is Doctor honoris causa of the Charles University in Prague, and the Military Academy in Liptovský Mikuláš. He is a member of the Scientific Society of Slovak Academy of Sciences.

Anna Tirpáková was born in Èadca, Slovakia, in April, 1954. She received her RNDr. degree in Theory of Mathematics Education at Comenius University in Bratislava in 1980, her PhD. degree in Probability and Mathematical Statistics in 1990 at Faculty of Mathematics, Physics and Informatics at Comenius University. She worked as a mathematics teacher at secondary school from 1977 to 1980, as a scientific worker from 1980-1990 at the Archaeological Institute of the Slovak Academy of Sciences, and since 1991 she has been a university professor at the Department of Mathematics of Faculty of Natural Sciences of Constantine the Philosopher University in Nitra.

Her research work is concentrated on probability theory and theory of fuzzy sets and their applications. She is an author of over 80 papers and co-author of several textbooks.

Renáta Bartková was born in Zvolen, Slovakia, in February, 1987. She received her PhD. degree in Probability and Mathematical Statistics in 2014 at Faculty of Natural Sciences at Matej Bel University in Banská Bystrica. She worked as a supporting specialist staff at the National Institute for Certified Educational

Measurements in Bratislava in 2015. Currently she works for Podravka International Ltd.

Her research work is concentrated on probability theory and theory of fuzzy sets and their applications in statistics. She is an author of three articles published in international journals.

Renáta Bartková, PhD. RNDr.
Podravka International s.r.o, Zvolen
Slovakia
Email: renata.hanesova@gmail.com

Beloslav Riečan, Dr.h.c. Prof. RNDr. DrSc.
Department of Mathematics
Faculty of Natural Sciences
Matej Bel University
Tajovského 40, SK-974 01 Banská Bystrica
Slovakia

Mathematical Institute Slovak Academy of Sciences
Štefánikova 49, SK-814 73 Bratislava
Slovakia
E-mail: Beloslav.Riecan@umb.sk

Anna Tirpáková, Prof. RNDr. CSc.
Department of Mathematics
Faculty of Natural Sciences
Constantine the Philosopher University
Trieda A. Hlinku 1, SK-949 74 Nitra
Slovakia
Email: atirpakova@ukf.sk

Address Correspondence to:

Prof. Beloslav Riečan,
Department of Mathematics
Faculty of Natural Sciences
Matej Bel University
Tajovského 40, 974 01 Banská Bystrica
Slovakia
E-mail: Beloslav.Riecan@umb.sk

CHAPTER 1

Kolmogorov Probability Theory

Abstract: Kolmogorov probability theory based on set theory belongs to the most important results of mathematics of the 20th century. Naturally, its main advantage is the possibility to use results of the modern measure theory. However, this fact sometimes does not allow larger considerations. In this chapter we want to show this paradox can be eliminated. Of course, we present only some basic ideas. Understanding them enables one to study further results and applications.

Keywords: Probability, Measure, Measurable functions, Random variable, Lebesgue integral, Independence, Limit theorems, Conditional expectation, Limit laws for maxima, Peaks over threshold.

1.1. PROBABILITY

Probability is a function of any set A from family S of sets which gives number $P(A) \in [0,1]$.

Definition 1.1.1 *Let Ω be a non-empty set, S be a family of subsets of Ω. We shall say that S is a σ-algebra if the following conditions are satisfied:*

(i) $\Omega \in S$,

(ii) $A \in S \Rightarrow A' = \Omega - A \in S$,

(iii) $A_n \in S \, (n = 1,2,\dots) \Rightarrow \bigcup_{n=1}^{\infty} A_n \in S$.

Of course, any σ-algebra is closed with respect to any usual operations, *i.e.*

$$\emptyset = \Omega - \Omega \in S.$$

If

$$A_1, \dots, A_n \in S,$$

then also

$$\bigcup_{i=1}^{n} A_i = A_1 \cup A_2 \cup \dots \cup A_n \cup \emptyset \cup \emptyset \dots \in S.$$

Similarly, if $A_n \in S \, (n = 1,2,\dots)$, then

Renáta Bartková, Beloslav Riečan and Anna Tirpáková

$$\bigcap_{n=1}^{\infty} A_n = \left(\bigcup_{n=1}^{\infty} A'_n \right)' \in \mathcal{S},$$

$$\bigcap_{i=1}^{n} A_i \in \mathcal{S}, n = 1,2,\dots,$$

or

$$A - B = A \cap B' \in \mathcal{S},$$

if $A, B \in \mathcal{S}$.

Definition 1.1.2 *Let $\mathcal{S}$ be a σ-algebra of subsets of a set Ω, $P: \mathcal{S} \to [0,1]$. We shall say that P is a probability, if the following properties are satisfied:*

(i) $P(\Omega) = 1$,

(ii) P is σ-additive, i.e.

$$P\left(\bigcup_{n=1}^{\infty} A_n \right) = \sum_{n=1}^{\infty} P(A_n),$$

if $A_n \in \mathcal{S}$, and $A_n \cap A_m = \emptyset (n \neq m)$.

Probably the simplest example of the probability space is a finite set $\Omega = \{\omega_1, \dots, \omega_n\}$ with the family $\mathcal{S}$ of all subsets of the set Ω and with the probability $P: \mathcal{S} \to [0,1]$ defined by the equality

$$P(A) = \frac{card\ A}{card\ \Omega},$$

where card A is the number of elements of the set A.

By the preceding the probability of one-element set $A = \Omega$ is the number

$$P(A) = \frac{1}{card\ \Omega} = \frac{1}{n}.$$

It seems to be a natural mathematical generalization of the model, where probabilities of the elements $\omega_1, \dots, \omega_n$ are non-negative numbers $p_1, \dots, p_n$ such that

$$\sum_{i=1}^{n} p_i = p_1 + p_2 + \dots + p_n = 1.$$

A special case is the choosing $p_i = \frac{1}{n}$ $(i = 1,2,\ldots,n)$. It is natural and effective (also with respect to the Kolmogorov theory) to define

$$P(A) = \sum_{x_i \in A} p_i = \sum \{p_i; x_i \in A\}.$$

It is easy to see that

$$P(\Omega) = \sum \{p_i; x_i \in \Omega\} = p_1 + p_2 + \ldots + p_n = 1.$$

From the axioms (i) and (ii) in Definition 1.1.2 some other properties can be derived.

Theorem 1.1.3 $P(\emptyset) = 0$.

Proof. Set $A_n = \emptyset$ $(n = 1,2,\ldots)$. The sets A_n belong to $\mathcal{S}$ $(i = 1,2,\ldots,n)$ and they are pairwise disjoint. Therefore

$$P(\emptyset) = P(\cup_{n=1}^{\infty} A_n) = \sum_{n=1}^{\infty} P(A_n) = \sum_{n=1}^{\infty} P(\emptyset) = \lim_{n \to \infty} \sum_{i=1}^{n} P(\emptyset) = \lim_{n \to \infty} nP(\emptyset).$$

The inequality $P(\emptyset) > 0$ leads to the equality $P(\emptyset) = \lim_{n \to \infty} nP(\emptyset) = \infty$, which is impossible. Therefore $P(\emptyset) = 0$.

Theorem 1.1.4 *Probability P is additive, i.e. if $A_1,\ldots,A_n \in \mathcal{S}, A_i \cap A_j = \emptyset (i \neq j)$, then*

$$P(\cup_{i=1}^{n} A_i) = \sum_{i=1}^{n} P(A_i).$$

Proof. Put $A_i = \emptyset$ for $i \geq n+1$. Then

$$P(\cup_{i=1}^{n} A_i) = P(\cup_{i=1}^{\infty} A_i) = \sum_{i=1}^{\infty} P(A_i) = \sum_{i=1}^{n} P(A_i).$$

Theorem 1.1.5 If $A \subset B$ $A, B \in \mathcal{S}$, then

$$P(B - A) = P(B) - P(A), P(A) \leq P(B).$$

In particular

$$P(A') = 1 - P(A).$$

Proof. Evidently

$$A, B - A \in \mathcal{S}, A \cap (B - A) = \emptyset.$$

Therefore

$$P(B) = P(A \cup (B - A)) = P(A) + P(B - A) \geq P(A).$$

If we put $B = \Omega$, then $\Omega - A = A'$, hence

$$1 = P(\Omega) = P(A) + P(A').$$

Remark 1.1.6 If $A_n \in \mathcal{S}, A_n \subset A_{n+1} (n = 1, 2, \ldots, n), A = \bigcup_{n=1}^{\infty} A_n$, then

$$P(A) = \lim_{n \to \infty} P(A_n).$$

Theorem 1.1.7 Let $A_n \in \mathcal{S}, A_n \subset A_{n+1} (n = 1, 2, \ldots, n), A = \bigcup_{n=1}^{\infty} A_n$

(we shall write $A_n \nearrow A$). Then

$$P(A) = \lim_{n \to \infty} P(A_n).$$

Proof. Put

$$B_1 = A_1, B_2 = A_2 - A_1, \ldots, B_n = A_n - A_{n-1}, \ldots$$

Then

$$B_n \in \mathcal{S} (i = 1, 2, \ldots, n), B_i \cap B_j = \emptyset (i \neq j),$$

and

$$\bigcup_{i=1}^{\infty} B_i = A.$$

Therefore

$$P(A) = \sum_{i=1}^{\infty} P(B_i) = \lim_{n \to \infty} \sum_{i=1}^{n} P(B_i).$$

But

$$\sum_{i=1}^{n} P(B_i) = P(\bigcup_{i=1}^{n} B_i) = P(A_n),$$

hence

$$P(A) = \lim_{n\to\infty} \sum_{i=1}^{n} P(B_i) = \lim_{n\to\infty} P(A_n).$$

Theorem 1.1.8 Let $A_n \in S, A_n \supset A_{n+1}(n = 1,2,\ldots,n), A = \bigcap_{n=1}^{\infty} A_n$

(we shall write $A_n \searrow A$). Then

$$P(A) = \lim_{n\to\infty} P(A_n).$$

Proof. Put

$$B_n = A'_n (n = 1,2,\ldots,n), B = A'.$$

Then

$$B_n = A'_n \nearrow A' = B.$$

Therefore

$$1 - P(A) = P(A') = P(B) = \lim_{n\to\infty} P(B_n) = \lim_{n\to\infty} P(A'_n) =$$

$$= \lim_{n\to\infty} P(1 - P(A_n)) = 1 - \lim_{n\to\infty} P(A_n),$$

hence

$$P(A) = \lim_{n\to\infty} P(A_n).$$

1.2. RANDOM VARIABLE

Definition 1.2.1 *Let (Ω, S, P) be a probability space, i.e. S is a σ-algebra of subsets of Ω, $P: S \to [0,1]$ is a probability. Random variable is a function $\xi: \Omega \to \mathbb{R}$ such that*

$$\{\omega \in \Omega;\ \xi(\omega) < x\} \in S$$

for every $x \in \mathbb{R}$. Distribution function F of a random variable ξ is the function $F: \mathcal{R} \to [0,1]$ defined by the equality

$$F(x) = P(\{\omega \in \Omega;\ \xi(\omega) < x\}).$$

Example 1.2.2 *Let $\Omega = \{\omega_1, \omega_2, \omega_3, \omega_4\}$, S be the family of all subsets of Ω, $P(\{\omega_1\}) = \frac{1}{2}$, $P(\{\omega_2\}) = \frac{1}{4}$, $P(\{\omega_3\}) = \frac{1}{8}$, $P(\{\omega_4\}) = \frac{1}{8}$.*

Further define

$$\xi(\omega_1) = 0, \xi(\omega_2) = -1, \xi(\omega_3) = 2, \xi(\omega_4) = 2.$$

If $x \leq -1$, then

$$\{\omega; \xi(\omega) < x\} = \{\omega; \xi(\omega) < -1\} = \emptyset,$$

hence $F(x) = P(\emptyset) = 0$. *If* $-1 < x \leq 0$, *then*

$$\{\omega; \xi(\omega) < x\} = \{\omega_2\},$$

hence $F(x) = P(\{\omega_2\}) = \frac{1}{4}$. *If* $0 < x \leq 2$, *then*

$$\{\omega; \xi(\omega) < x\} = \{\omega_1, \omega_2\},$$

hence $F(x) = P(\{\omega_1\}) + P(\{\omega_2\}) = \frac{1}{2} + \frac{1}{4} = \frac{3}{4}$. *If* $2 \leq x$, *then*

$$\{\omega; \xi(\omega) < x\} = \{\omega_1, \omega_2 \, \omega_3, \omega_4\},$$

hence $F(x) = P(\Omega) = 1$.

Theorem 1.2.3 *Let F be the distribution function of a random variable ξ. Then the following properties are satisfied:*

(i) F is non-decreasing;

(ii) F is left continuous in any point;

(iii) $\lim_{x \to \infty} F(x) = 1$;

(iv) $\lim_{x \to -\infty} F(x) = 0$.

Proof. If $x_1 < x_2$, then

$$\{\omega; \xi(\omega) < x_1\} \subset \{\omega; \xi(\omega) < x_2\},$$

hence

$$F(x_1) = P(\{\omega; \xi(\omega) < x_1\}) \leq P(\{\omega; \xi(\omega) < x_2\}) = F(x_2),$$

and F is non-decreasing. Let $x_0 \in \mathbb{R}$ be arbitrary. Let $x_n \nearrow x_0$ be an arbitrary sequence. We have to prove $\lim_{n \to -\infty} F(x_n) = F(x_0)$. But

$$A_n = \{\omega; \ \xi(\omega) < x_n\} \nearrow A = \{\omega; \ \xi(\omega) < x_0\}.$$

Therefore

$$F(x_0) = P(A) = \lim_{n \to -\infty} P(A_n) = \lim_{n \to -\infty} F(x_n).$$

Further let $\{x_n\}_n$ be an arbitrary non-decreasing sequence such that

$\lim_{n \to \infty} x_n = \infty$. Then

$$1 = P(\Omega) = P(\cup_{n=1}^{\infty} \{\omega; \ \xi(\omega) < x_n\}) = \lim_{n \to \infty} P(\{\omega; \ \xi(\omega) < x_n\}) = \lim_{n \to \infty} F(x_n).$$

Therefore

$$\lim_{x \to \infty} F(x) = 1.$$

Finally, let $\{x_n\}_n$ be a non-increasing sequence, $\lim_{n \to \infty} x_n = -\infty$. Then

$$0 = P(\emptyset) = P(\cap_{n=1}^{\infty} \{\omega; \ \xi(\omega) < x_n\}) = \lim_{n \to \infty} P(\{\omega; \ \xi(\omega) < x_n\}) = \lim_{n \to \infty} F(x_n).$$

Therefore

$$\lim_{x \to -\infty} F(x) = 0.$$

Because a random variable is a mapping $\xi \colon \Omega \to \mathbb{R}$, we can use also the notion of the pre-image of $A \subset \mathbb{R}$, what is the set

$$\xi^{-1}(A) = \{\omega \in \Omega; \ \xi(\omega) \in A\}.$$

With respect to the notation we have

$$F(x) = P(\xi^{-1}((-\infty, x))).$$

But not only pre-images of intervals of the type $(-\infty, x)$ belongs to $\mathcal{S}$.

Theorem 1.2.4 Let $\xi \colon \Omega \to \mathbb{R}$ *be a random variable,* $\mathcal{T} = \{A \subset \mathbb{R}; \ \xi^{-1}(A) \in \mathcal{S}\}$. *Then* $\mathcal{T}$ *is a σ-algebra of subsets of the set* $\mathbb{R}$.

Proof. First

$$\xi^{-1}(\mathbb{R}) = \Omega \in \mathcal{S},$$

hence $\mathbb{R} \in \mathcal{T}$. Further, let $A \in \mathcal{T}$, *i.e.* $\xi^{-1}(A) \in \mathcal{S}$. Then

$$\xi^{-1}(A') = (\xi^{-1}(A)) \in \mathcal{S},$$

hence $A' \in \mathcal{T}$. Finally, let $A_n \in \mathcal{T}$ $(n = 1,2,\ldots,n)$, *i.e.* $\xi^{-1}(A_n) \in \mathcal{S}$ $(n = 1,2,\ldots)$. Then

$$\xi^{-1}(\textstyle\bigcup_{n=1}^{\infty} A_n) = \textstyle\bigcup_{n=1}^{\infty} \xi^{-1}(A_n) \in \mathcal{S},$$

hence

$$\textstyle\bigcup_{n=1}^{\infty} A_n \in \mathcal{T}.$$

Of course, there are many σ-algebras similar to the family $\mathcal{T}$ from Theorem 2.4, *e.g.* the family of all subsets of $\mathbb{R}$ has the property. We shall use the intersection of all σ-algebras containing the family $\mathcal{J}$ of all intervals of of the type $(-\infty, x)$. Generally under arbitrary non-empty family $\mathcal{P}$ of subsets of $\mathcal{R}$ there exists the σ-algebra containing $\mathcal{P}$.

Theorem 1.2.5 *Let $\mathcal{P}$ be an arbitrary non-empty family of subsets of the set $\mathbb{R}$. Then there exists the σ-algebra $\sigma(\mathcal{P}) \supset \mathcal{P}$ such that $\sigma(\mathcal{P}) \subset \mathcal{U}$ for every σ-algebra $\mathcal{U}$ such that $\mathcal{P} \subset \mathcal{U}$. The σ-algebra $\sigma(\mathcal{P})$ is determined uniquely.*

Proof. Let $\sigma(\mathcal{P})$ be the intersection of all σ-algebras containing the family $\mathcal{P}$. Let $A \in \sigma(\mathcal{P})$. Since $\mathcal{P} \subset \mathcal{U}$, also $\sigma(\mathcal{P}) \subset \mathcal{U}$, hence $A \in \mathcal{U}$. Since $\mathcal{U}$ is a σ-algebra, also $A' \in \mathcal{U}$. Since $A' \in \mathcal{U}$ for any σ-algebra containing $\mathcal{P}$, we have $A' \in \sigma(\mathcal{P})$. Finally, let

$$A_n \in \sigma(\mathcal{P})(n = 1,2,\ldots),$$

and $\mathcal{U}$ be any σ-algebra containing $\mathcal{P}$. Then

$$A_n \in \mathcal{U}(n = 1,2,\ldots),$$

hence

$$\textstyle\bigcup_{n=1}^{\infty} A_n \in \mathcal{U}.$$

Since $\mathcal{U}$ is any σ-algebra over $\mathcal{P}$, we have

$$\bigcup_{n=1}^{\infty} A_n \in \sigma(\mathcal{P}).$$

Definition 1.2.6 *Let $\mathcal{J}$ be the family of all intervals of the type $(-\infty, x)$, where $x \in \mathbb{R}$. Then the σ-algebra $\sigma(\mathcal{J})$ will be denoted by $\mathcal{B}(\mathbb{R})$ and the sets belonging to the σ-algebra $\mathcal{B}(\mathbb{R})$ will be called Borelian.*

Theorem 1.2.7 *Let $\xi \colon \Omega \to \mathbb{R}$ be a random variable. Then for any set*

$A \in \mathcal{B}(\mathbb{R})$ there is $\xi^{-1}(\mathcal{A}) \in \mathcal{S}$.

Proof. Let $\mathcal{J} = \{(-\infty, x); \ x \in \mathbb{R}\}$, $\mathcal{U} = \{A \subset \mathbb{R}; \ \xi^{-1}(\mathcal{A}) \in \mathcal{S}\}$. Evidently $\mathcal{U} \supset \sigma(\mathcal{J})$. By Theorem 2.4 the family $\mathcal{U}$ is a σ-algebra. Therefore

$$\mathcal{U} \supset \sigma(\mathcal{J}) = \mathcal{B}(\mathbb{R})$$

Hence, if $A \in \mathcal{B}(\mathbb{R})$, then $A \in \mathcal{U}$, hence $\xi^{-1}(\mathcal{A}) \in \mathcal{S}$.

Of course, Borel sets can be characterized by various ways. We shall present it in one example.

Example 1.2.8 Let $\mathcal{C} = \{[a, b]; a, b \in \mathbb{R}\}$. Then $\sigma(\mathcal{C}) = \mathcal{B}(\mathbb{R})$. We have to show two inclusions. First let $[a, b] \in \mathcal{C}$. We know that $\left(-\infty; b + \frac{1}{n}\right) \in \mathcal{J} \subset \mathcal{B}(\mathbb{R})$ for any $n \in N$. Therefore also

$$\left[a, b + \frac{1}{n}\right) = \left(-\infty, b + \frac{1}{n}\right) - (-\infty, a) \in \mathcal{B}(\mathbb{R}).$$

We have

$$[a, b] = \bigcap_{n=1}^{\infty} \left[a, b + \frac{1}{n}\right) \in \mathcal{B}(\mathbb{R}),$$

hence

$$\mathcal{C} \subset \mathcal{B}(\mathbb{R}).$$

Since $\mathcal{B}(\mathbb{R})$ is a σ-algebra, we have $\sigma(\mathcal{C}) \subset \mathcal{B}(\mathbb{R})$. In other hand, let $(-\infty, c) \in \mathcal{J}$. Then

$$(-\infty, c) = \bigcup_{n=1}^{\infty} \left[c - n, c - \frac{1}{n} \right] \in \sigma(\mathcal{C}),$$

hence

$$\mathcal{J} \subset \sigma(\mathcal{C}).$$

Since $\sigma(\mathcal{C})$ is a σ-algebra, we have

$$\mathcal{B}(\mathbb{R}) = \sigma(\mathcal{J}) \subset \sigma(\mathcal{C}).$$

Any random variable can be characterized by its distribution function. Moreover also by a probability on $\mathcal{B}(\mathbb{R})$.

Theorem 1.2.9 *Let $\xi \colon \Omega \to \mathbb{R}$ be a random variable, $F \colon \mathbb{R} \to [0,1]$ be its distribution function. Then there exists exactly one probability $\lambda_F \colon \mathcal{B}(\mathbb{R}) \to [0,1]$ such that*

$$\lambda_F([a, b)) = F(b) - F(a)$$

for all intervals $[a, b)$.

Proof.

1. Existence. For any $A \in \mathcal{B}(\mathbb{R})$ put $\lambda_F(A) = P(\xi^{-1}(A))$. We have

$$\lambda_F(\mathbb{R}) = P(\xi^{-1}(\mathbb{R}))) = P(\Omega) = 1.$$

Let $A_n \in \mathcal{B}(\mathbb{R})$ $(n = 1,2,\ldots,n), A_i \cap A_j = \emptyset$ $(i \neq j)$. Then

$$\lambda_F(\textstyle\bigcup_{n=1}^{\infty} A_n) = P\big(\xi^{-1}(\bigcup_{n=1}^{\infty} A_n)\big) = P(\bigcup_{n=1}^{\infty} \big(\xi^{-1}(A_n)\big).$$

But for $i \neq j$

$$\xi^{-1}(A_i) \cap \xi^{-1}(A_j) = \xi^{-1}(A_i \cap A_j) = \xi^{-1}(\emptyset) = \emptyset.$$

Therefore

$$\lambda_F(\textstyle\bigcup_{n=1}^{\infty} A_n) = P\big(\bigcup_{n=1}^{\infty} \xi^{-1}(A_n)\big) = \sum_{n=1}^{\infty} P\big(\xi^{-1}(A_n)\big) = \sum_{n=1}^{\infty} \lambda_F(A_n).$$

Moreover

$$\lambda_F([a, b)) = \lambda_F((-\infty, b)) - \lambda_F((-\infty, a)) =$$

$$= P(\xi^{-1}((-\infty, b))) - P(\xi^{-1}((-\infty, a))) = F(b) - F(a).$$

2. Uniqueness. Let $\mu: \mathcal{B}(\mathbb{R}) \rightarrow [0,1]$ be a probability such that

$$\mu([a, b)) = F(b) - F(a)$$

for all intervals [a,b). Define the family of sets

$$\mathcal{K} = \{A \in \mathcal{B}(\mathbb{R}); \ \mu(A) = \lambda_F(A)\}.$$

By Theorems 1.7 and 1.8 it follows that $\mathcal{K}$ is monotone, *i.e.*

$$A_n \in \mathcal{K}(n = 1,2,\dots), A_n \nearrow A \Rightarrow A \in \mathcal{K},$$

$$B_n \in \mathcal{K}(n = 1,2,\dots), B_n \searrow B \Rightarrow B \in \mathcal{K}.$$

Moreover $\mathcal{K} \supset \mathcal{R}$, where $\mathcal{R}$ consists of finite unions of intervals of the type $[a, b)$ and intervals of the type $(-\infty, c)$. Let $M(\mathcal{R})$ be the least monotone family over $\mathcal{R}$. By the definition $\mathcal{K} \supset M(\mathcal{R})$. If we will show that $M(\mathcal{R})$ is an algebra, we obtain that $M(\mathcal{R})$ is a σ-algebra over $\mathcal{J}$, hence $\mathcal{K} \supset \sigma(\mathcal{J}) = \mathcal{B}(\mathcal{R})$. Consider fixed $A \in \mathcal{R}$ and put

$$\mathcal{K}_A = \{B \in M(\mathcal{R}); \ A \cup B \in M(\mathcal{R})\}.$$

By the assumption $\mathcal{K} \supset \mathcal{R}$. It is easy see that $\mathcal{K}_A$ is monotone, hence $A \cup B \in M(\mathcal{R})$ for all $B \in M(\mathcal{R})$. Choose now fixed $B \in M(\mathcal{R})$ and define

$$\mathcal{L}_B = \{A \in M(\mathcal{R}); \ A \cup B \in M(\mathcal{R})\}.$$

By previous $\mathcal{L}_B \supset \mathcal{R}$. Since $\mathcal{L}_B$ is monotone, we have $\mathcal{L}_B \supset M(\mathcal{R})$, hence $A \in \mathcal{L}_B$ for any $A \in M(\mathcal{R})$, hence $A \cup B \in M(\mathcal{R})$. We have proved that $M(\mathcal{R})$ is closed with respect to unions. Similarly it can be proved that $M(\mathcal{R})$ is closed with respect to differences of sets. Moreover

$$\mathbb{R} = \bigcup_{n=1}^{\infty} ((-\infty, n)) \in M(\mathcal{R}),$$

hence $\mathcal{M}(\mathcal{R})$ is really a σ-algebra.

1.3. MEAN VALUE

The subject of the probability theory is randomness. But the Kolmogorov axiomatic is based on the measure theory [93]. There are three basic notions:

    1. probability=measure,

    2. random variable=measurable function,

    3. mean value=integral.

Let us give a motivation of the mean value of a random variable ξ as the integral

$$E(\xi) = \int_\Omega \xi dP.$$

Let ξ have a finite number of values

$$\alpha_1, \alpha_2, \ldots, \alpha_n$$

with probabilities

$$p_1, p_2, \ldots, p_n,$$

hence

$$p_i = P(\{\omega \in \Omega; \ \xi(\omega) = \alpha_i\}) = P(\xi^{-1}(\{\alpha_i\})).$$

But

$$\sum_{i=1}^n \alpha_i p_i = \sum_{i=1}^n \alpha_i P(\{\omega \in \Omega; \ \xi(\omega) = \alpha_i\}) = \int_\Omega \xi dP.$$

To define the integral of the function by such a way is usual in the modern mathematics.

Definition 1.3.1 *Random variable $\xi: \Omega \to \mathbb{R}$ is simple, if it has a finite number of values:*

$$\alpha_1, \alpha_2, \ldots, \alpha_n.$$

Its mean value is defined as the integral

$$\int_\Omega \xi dP = \sum_{i=1}^n \alpha_i P(\{\omega \in \Omega; \ \xi(\omega) = \alpha_i\}).$$

If ξ is non-negative random variable, we consider a sequence $(\xi_n)_n$ of non-negative simple functions such that $\xi_n \nearrow \xi$. The function ξ is integrable if limit $\lim_{n\to\infty} \int_\Omega \xi_n dP$ is finite. In the case define

$$\int_\Omega \xi dP = \lim_{n\to\infty} \int_\Omega \xi_n dP.$$

Finally, an arbitrary random variable $\xi: \Omega \to \mathbb{R}$ is integrable, if the following non-negative functions are integrable

$$\xi^+ = \max(\xi, 0), \xi^- = \max(-\xi, 0).$$

In the case we define

$$\int_\Omega \xi dP = \int_\Omega \xi^+ dP - \int_\Omega \xi^- dP.$$

Of course, for the correctness of the Definition 3.1 it is necessary to prove the following properties:

1. To any $\xi \geq 0$ there exists a non-decreasing sequence $(\xi_n)_n$ of simple functions such that $\xi = \lim_{n\to\infty} \xi_n, \xi_n \nearrow \xi$.

2. The number $\lim_{n\to\infty} \int_\Omega \xi_n dP$ does not depend on the choice of the sequence $(\xi_n)_n$, it depends only on the function ξ.

Remark 1.3.2 *Increasing sequence of function is defined in [1], [92].*

Theorem 1.3.3 *To any non-negative random variable ξ there exists a sequence $(\xi_n)_n$ of non-negative functions such that $\xi_n \nearrow \xi$.*

Proof. Put

$$\xi_n(\omega) = n,$$

if

$$\xi(\omega) \geq n,$$

and

$$\xi_n(\omega) = \frac{j-1}{2^n},$$

if

$$\frac{j-1}{2^n} \leq \xi(\omega) < \frac{j}{2^n}$$

$(n = 1, 2, \ldots, n2^n)$ resp.

Theorem 1.3.4 *Let ξ be a non-negative random variable, $(\xi_n)_n$, $(\eta_m)_m$ are sequences of simple, non-negative functions such that $\xi_n \nearrow \xi$, $\eta_m \nearrow \xi$. Then*

$$\lim_{n \to \infty} \int_\Omega \xi_n dP = \lim_{m \to \infty} \int_\Omega \eta_m dP.$$

In the proof of Theorem 3.4 two lemmas will be used.

Lemma 1.3.5 *Let $(\xi_n)_n$ be a sequence of simple random variables such that $\xi_n \searrow 0$. Then*

$$\lim_{n \to \infty} \int_\Omega \xi_n dP = 0.$$

Proof. Let $\varepsilon > 0$ be an arbitrary number. Define

$$A_n = \{\omega \in \Omega; \xi_n \geq \varepsilon\}.$$

Then

$$\int_\Omega \xi_n dP = \int_\Omega \xi_n \chi_{A_n} dP + \int_\Omega \xi_n \chi_{A'_n} dP \leq P(A_n)\max\xi_1 + \varepsilon.$$

But $A_n \searrow 0$, hence $\lim_{n \to \infty} P(A_n) = 0$ by Theorem 1.8. Therefore

$$\lim_{n \to \infty} \int_\Omega \xi_n dP \leq \varepsilon$$

for any $\varepsilon > 0$, hence

$$\lim_{n \to \infty} \int_\Omega \xi_n dP = 0.$$

Lemma 1.3.6 Let η_m be simple, η simple, $\eta_m \nearrow \eta$. Then

$$\lim_{m \to \infty} \int_\Omega \eta_m dP = \int_\Omega \eta dP.$$

Proof. It is sufficient to consider $\xi_n = \eta - \eta_m$ and to use Lemma 3.5.

Proof. Proof of Theorem 3.4. Let $0 \le \xi_n \nearrow \xi$, $0 \le \eta_m \nearrow \xi$, ξ_n, η_m be simple. For a fixed m

$$\min(\xi_n, \eta_m) \nearrow \min(\xi, \eta_m) = \eta_m.$$

Therefore by Lemma 3.6

$$\int_\Omega \eta_m dP = \lim_{n \to \infty} \int_\Omega \min(\xi_n, \eta_m) dP \le \lim_{n \to \infty} \int_\Omega \xi_n dP.$$

By the preceding inequality one can obtain

$$\lim_{m \to \infty} \int_\Omega \eta_m dP \le \lim_{n \to \infty} \int_\Omega \xi_n dP.$$

Since the mean value has been defined as the abstract integral, we can use usual theorems holding in integration theory. *E.g.* ξ, η are integrable, and $\alpha, \beta \in \mathbb{R}$, then

$$E(\alpha \xi + \beta \eta) = \alpha E(\xi) + \beta E(\eta).$$

Hence, *e.g.* if $\xi_1, \ldots, \xi_n$ have same mean value $E(\xi_i) = a$, and $\overline{\xi}$ is their arithmetic average,

$$\overline{\xi} = \frac{1}{n} \sum_{i=1}^{n} \xi_i,$$

then

$$E(\overline{\xi}) = E\left(\frac{1}{n} \sum_{i=1}^{n} \xi_i\right) = \frac{1}{n} E\left(\sum_{i=1}^{n} \xi_i\right) = \frac{1}{n} \sum_{i=1}^{n} E(\xi_i) = \frac{1}{n} na = a.$$

Less known is the theorem about monotone convergence: If $\xi_n \nearrow \xi$, ξ_n are integrable, and $\lim_{n \to \infty} E(\xi_n) < \infty$, then ξ is integrable, and

$$E(\xi) = \lim_{n \to \infty} E(\xi_n).$$

Another version for infinite series. If $\xi_n \ge 0$ are integrable, and $\sum_{i=1}^{\infty} E(\xi_n) < \infty$, then

$$E(\textstyle\sum_{n=1}^{\infty} \xi_n) = \sum_{n=1}^{\infty} E(\xi_n).$$

1.4. TRANSFORMATION OF INTEGRAL

Of course, the usual theorems for integration are in Euclidean space. We have defined mean value with respect to the given σ-algebra S of subsets of Ω, and a given probability $P: S \to [0,1]$. In the set $\mathbb{R}$ of real numbers we have the σ-algebra $\mathcal{B}(\mathbb{R})$ and on it (with respect to a random variable ξ with the distribution function F) the probability $\lambda_F: \mathcal{B}(\mathbb{R}) \to [0,1]$. We shall consider the composed function $g \circ \xi$, where $g: \mathbb{R} \to \mathbb{R}$. If $g(x) = x$, then $g \circ \xi = \xi$, hence

$$E(g \circ \xi) = E(\xi).$$

Another useful example gives the function $g(x) = (x - E(\xi))^2$ what leads to the dispersion

$$E(g \circ \xi) = E(\xi - E(\xi))^2 = D(\xi).$$

The same way we can obtain a transformation formula for a larger class of functions $g: \mathbb{R} \to \mathbb{R}$.

Definition 1.4.1 *The function $g: \mathbb{R} \to \mathbb{R}$ is called Borel, if*

$$A \in \mathcal{B}(\mathbb{R}) \Rightarrow g^{-1}(A) \in \mathcal{B}(\mathbb{R}).$$

Example 1.4.2 *Every continuous function is Borel. Namely, it is easy to prove that $\mathcal{B}(\mathbb{R}) = \sigma(\mathcal{O})$, where $\mathcal{O}$ is the family of all open intervals. Let g be continuous, and*

$$\mathcal{K} = \{A \in \mathcal{B}(\mathbb{R}); g^{-1}(A) \in \mathcal{B}(\mathbb{R})\}.$$

Then $\mathcal{K} \supset \mathcal{O}$, and $\mathcal{K}$ is a σ-algebra, hence $\mathcal{K} \supset \mathcal{O} = \mathcal{B}(\mathbb{R})$.

Theorem 1.4.3 *Let $\xi: \Omega \to \mathbb{R}$ be a random variable, $g: \mathbb{R} \to \mathbb{R}$ be a Borel function. If $g \circ \xi: \Omega \to \mathbb{R}$ is integrable, then g is integrable with respect to λ_F. In the case equality*

$$\int_{\Omega} g \circ \xi dP = \int_{\Omega} g d\lambda_F$$

holds.

Example 1.4.4 *If $g(x) = x$, then $g \circ \xi = \xi$, hence*

$$E(\xi) = \int_\Omega g \circ \xi \, dP = \int_\mathbb{R} g \, d\lambda_F = \int_{-\infty}^\infty x \, d\lambda_F(x).$$

Example 1.4.5 *If $g(x) = (x - E(\xi))^2$, then $g \circ \xi = (\xi - E(\xi))^2$, hence*

$$D(\xi) = E((\xi - E(\xi))^2) = E(g \circ \xi) = \int_\mathbb{R} g(x) \, d\lambda_F(x) = \int_{-\infty}^\infty (x - E(\xi))^2 \, d\lambda_F(x).$$

Proof. Proof of Theorem 1.4.3. Let g be simple, $g = \sum_{i=1}^n \alpha_i \chi_{A_i}$, $\alpha_i \in \mathbb{R}$, $A_i \in \mathcal{B}(\mathbb{R})$, A_i disjoint. Then

$$g(\xi(\omega)) = \sum_{i=1}^n \alpha_i \chi_{\xi^{-1}(A_i)}(\omega),$$

hence

$$\int_\Omega g \circ \xi \, dP = \sum_{i=1}^n \alpha_i P(\xi^{-1}(A_i)) = \sum_{i=1}^n \alpha_i \lambda_F(A_i) = \int_{-\infty}^\infty g(x) \, d\lambda_F(x).$$

If $g \geq 0$, and g_n are simple, $g_n \nearrow g$, then $g_n \circ \xi \nearrow g \circ \xi$, hence

$$\int_\Omega g \circ \xi \, dP = \lim_{n \to \infty} \int_\Omega g_n \circ \xi \, dP = \lim_{n \to \infty} \int_\mathbb{R} g_n \, d\lambda_F = \int_\mathbb{R} g \, d\lambda_F.$$

Finally in the general case

$$g = g^+ - g^-,$$

hence

$$g \circ \xi = g^+ \circ \xi - g^- \circ \xi$$

and

$$\int_\Omega g \circ \xi \, dP = \int_\Omega g^+ \circ \xi \, dP - \int_\Omega g^- \circ \xi \, dP = \int_\mathbb{R} g^+ \, d\lambda_F - \int_\mathbb{R} g^- \, d\lambda_F = \int_\mathbb{R} g \, d\lambda_F.$$

It remains present the methods of counting of the mean value in usual cases of discrete, or continuous random variables, resp.

Theorem 1.4.6 *Let $\xi : \Omega \to \mathbb{R}$ be a random variable having values $x_1, \ldots, x_n$ with probabilities $p_1, \ldots, p_n$. Then*

$$E(g \circ \xi) = \sum_{i=1}^n g(x_i) p_i.$$

Proof. Put $A = \{x_1, \ldots, x_n\}$. Since $\lambda_F(\mathbb{R} - A) = 0$, we have

$$E(g \circ \xi) = \int_{\mathbb{R}} g d\lambda_F = \int_{\mathbb{R}} \chi_A g d\lambda_F = \sum_{i=1}^{n} \chi_{\{x_i\}}(x_i) g(x_i) d\lambda_F(\{x_i\}) = \sum_{i=1}^{n} g(x_i) p_i.$$

Theorem 1.4.7 *Let $\xi: \Omega \to \mathbb{R}$ be a random variable with the density f, i.e.*

$$P(\xi < x) = \int_{-\infty}^{x} f(t) dt, x \in \mathbb{R}. \, Then$$

$$E(g \circ \xi) = \int_{-\infty}^{\infty} g(x) f(x) dx.$$

Proof. Let $g = \sum_{i=1}^{n} \alpha_i \chi_{(A_i)}$. Then

$$E(g \circ \xi) = \int_{\mathbb{R}} g d\lambda_F = \sum_{i=1}^{n} \alpha_i \lambda_F(A_i) = \sum_{i=1}^{n} \alpha_i \int_{\mathbb{R}} \chi_{A_i}(x) f(x) dx =$$

$$= \int_{\mathbb{R}} \left(\sum_{i=1}^{n} \alpha_i \chi_{A_i}(x) \right) f(x) dx = \int_{-\infty}^{x} g(x) f(x) dx.$$

If g_n are non-negative simple, and $g_n \nearrow g$, then also $g_n f \nearrow g f$, hence

$$\int_{\mathbb{R}} g(x) f(x) dx = \lim_{n \to \infty} \int_{\mathbb{R}} g_n(x) f(x) dx = \lim_{n \to \infty} \int_{\mathbb{R}} g_n d\lambda_F = \int_{\mathbb{R}} g d\lambda_F = E(g \circ \xi).$$

Finally, for an arbitrary g we have

$$E(g \circ \xi) = E(g^+ \circ \xi) - E(g^- \circ \xi) = \int_{\mathbb{R}} g^+(x) f(x) dx - \int_{\mathbb{R}} g^-(x) f(x) dx =$$

$$= \int_{\mathbb{R}} g(x) f(x) dx.$$

Really we have in the discrete case *e.g.*

$$E(\xi) = \sum_{i=1}^{n} x_i p_i, D(\xi) = \sum_{i=1}^{n} (x_i - E(\xi))^2 p_i,$$

and in the continuous case

$$E(\xi) = \int_{-\infty}^{\infty} x f(x) dx, D(\xi) = \int_{-\infty}^{\infty} (x - E(\xi))^2 f(x) dx.$$

Example 1.4.8 *Let $x_i = i$, $p_i = e^{-\lambda} \frac{\lambda^i}{i!} (i = 1, 2, \ldots)$ (the Poisson distribution). Then*

$$E(\xi) = \sum_{i=1}^{\infty} i e^{-\lambda} \frac{\lambda^i}{i!} = \lambda e^{-\lambda} \sum_{i=1}^{\infty} \frac{\lambda^{i-1}}{(i-1)!} = \lambda.$$

Example 1.4.9 *Let $f(x) = 0$, if $x \leq 0$, $f(x) = \lambda e^{-\lambda x}$, if $x > 0$ (the exponential distribution). Then*

$$E(\xi) = \int_0^\infty x\lambda e^{-\lambda x} d(x) = \left[x\lambda \frac{e^{-\lambda x}}{-\lambda} \right]_0^\infty - \int_0^\infty 1\lambda \frac{e^{-\lambda x}}{-\lambda} d(x) = \left[\frac{e^{-\lambda x}}{-\lambda} \right]_0^\infty = \frac{1}{\lambda}.$$

In the dispersion it is useful to count the number $(\xi - E(\xi))^2$, hence

$$D(\xi) = E(\xi^2 - 2\xi E(\xi) + E(\xi)^2) = E(\xi^2) - 2E(\xi)E(\xi) + E((E(\xi))^2) =$$

$$= E(\xi^2) - E(\xi)^2.$$

1.5. INDEPENDENCE

A classical example is a repetition of experiences. *E.g.* if we loss two-times cube, and A means, that in the first experience we have odd number, and B means that in the second experience we obtain the number 6. The experiment can be described by the help of all ordered pairs of numbers $1, 2, \ldots, 6$,

$$\Omega = \{(1,1), (1,2), \ldots, (1,6), \ldots, (6,6)\}.$$

Then

$$A = \{(1,1), (1,2), \ldots, (1,6), (3,1), \ldots, (3,6), (5,1), \ldots, (5,6)\},$$

$$B = \{(1,6), (2,6), (3,6), \ldots, (6,6)\},$$

$$A \cap B = \{(1,6), (3,6), (5,6)\},$$

hence

$$P(A) = \frac{card A}{card \Omega} = \frac{3 \cdot 6}{6 \cdot 6} = \frac{3}{6} = \frac{1}{2},$$

$$P(B) = \frac{card B}{card \Omega} = \frac{6}{6 \cdot 6} = \frac{1}{6},$$

then

$$P(A \cap B) = \frac{3}{6 \cdot 6} = \frac{1}{2} \cdot \frac{1}{6} = P(A)P(B).$$

This leads *e.g.* to the well known formula

$$P_{n,k} = n_k \; p^k(1-p)^{n-k}.$$

Definition 1.5.1 *Random variables ξ, η are independent if for any $A, B \in \mathcal{B}(\mathbb{R})$ the following equality holds*

$$P(\xi^{-1}(A) \cap \eta^{-1}(B)) = P(\xi^{-1}(A)).P(\eta^{-1}(B)).$$

Theorem 1.5.2 *Random variables ξ, η are independent if and only if*

$$P(\xi^{-1}((-\infty, x)) \cap \eta^{-1}((-\infty, y))) = P(\xi^{-1}((-\infty, x))).P(\eta^{-1}((-\infty, y)))$$

for any $x, y \in \mathbb{R}$.

Proof. If ξ, η are independent, it is sufficient to put $A = (-\infty, x)$, $B = (-\infty, y)$. On the other hand, let equality mentioned in the theorem holds. Take $(-\infty, y)$ fixed and consider the family

$$\mathcal{K} = \{A \in \mathcal{B}(\mathbb{R}); P(\xi^{-1}(A) \cap \eta^{-1}((-\infty, y))) = P(\xi^{-1}(A)).P(\eta^{-1}((-\infty, y)))\}.$$

The family $\mathcal{K}$ is monotone, and $\mathcal{K} \supset \mathcal{J}$, hence $\mathcal{K} \supset \mathcal{M}(\mathcal{J}) = \mathcal{B}(\mathbb{R})$, and

$$P(\xi^{-1}(A) \cap \eta^{-1}((-\infty, y))) = P(\xi^{-1}(A)).P(\eta^{-1}((-\infty, y)))$$

for all $A \in \mathcal{B}(\mathbb{R})$. Now for fixed $A \in \mathcal{B}(\mathbb{R})$ put

$$\mathcal{L} = \{B \in \mathcal{B}(\mathbb{R}); P(\xi^{-1}(A) \cap \eta^{-1}(B)) = P(\xi^{-1}(A)).P(\eta^{-1}(B))\}.$$

By the preceding $\mathcal{L} \supset \mathcal{J}$. Since $\mathcal{L}$ is monotone, we have $\mathcal{L} \supset \mathcal{M}(\mathcal{J}) = \mathcal{B}(\mathbb{R})$, hence the equality stated in Definition 5.1 holds for any $A, B \in \mathcal{B}(\mathbb{R})$.

Theorem 1.5.3 *If ξ, η are independent and integrable, then $\xi\eta$ is integrable and*

$$E(\xi\eta) = E(\xi)E(\eta).$$

Proof. Let ξ, η be simple,

$$\xi = \sum_{i=1}^{n} \alpha_i \chi_{A_i}(x), \eta = \sum_{j=1}^{m} \beta_j \chi_{B_j}(x).$$

Then

$$\xi\eta = \sum_{i=1}^{n}\sum_{j=1}^{m}\alpha_i\beta_j\chi_{A_i \cap B_j},$$

hence

$$E(\xi\eta) = \sum_{i=1}^{n}\sum_{j=1}^{m}\alpha_i\beta_j P(A_i \cap B_j) = \sum_{i=1}^{n}\sum_{j=1}^{m}\alpha_i\beta_j P(A_i) \cdot P(B_j) =$$

$$= \left(\sum_{i=1}^{n}\alpha_i P(A_i)\right)\left(\sum_{j=1}^{m}\beta_j P(B_j)\right) = E(\xi)E(\eta).$$

If ξ, η are non-negative, we can use $0 \le \xi_n \nearrow \xi$, $0 \le \eta_m \nearrow \eta$, ξ_n, η_n simple (Theorem 3.3). Then $\xi_n \cdot \eta_m \nearrow \xi \cdot \eta$, hence

$$E(\xi\eta) = \lim_{n\to\infty}\lim_{m\to\infty}E(\xi_n \cdot \eta_m) = \lim_{n\to\infty}\lim_{m\to\infty}E(\xi_n)E(\eta_m) =$$

$$= \lim_{n\to\infty}E(\xi_n)\lim_{m\to\infty}E(\eta_m) = E(\xi)E(\eta).$$

In the general case we have $\xi = \xi^+ - \xi^-$, $\eta = \eta^+ - \eta^-$. The functions ξ^+, ξ^- on the first side, or η^+, η^- on the second side, resp., are pairwise independent. Namely

$$(\xi^+)^{-1}(A) = \xi^{-1}(A) \cap [0,\infty)),$$

$$(\xi^-)^{-1}(A) = \xi^{-1}(-A) \cap [0,\infty)),$$

and similarly η^+, η^-. Therefore

$$E(\xi\eta) = E\big((\xi^+ - \xi^-)(\eta^+ - \eta^-)\big) =$$

$$= E(\xi^+\eta^+) - E(\xi^-\eta^+) - E(\xi^-\eta^+) - E(\xi^-\eta^-) =$$

$$= E(\xi^+)E(\eta^+) - E(\xi^-)E(\eta^+) - E(\xi^-)E(\eta^+) - E(\xi^-)E(\eta^-) =$$

$$= E((\xi^+) - E(\xi^-))(E(\eta^+) - E(\eta^-)).$$

Theorem 1.5.4 *Let ξ, η be independent and they have integrable quadrants ξ^2, η^2. Then $\xi + \eta$ has integrable quadrant $(\xi + \eta)^2$, and*

$$D(\xi + \eta) = D(\xi) + D(\eta).$$

Proof. We have

$$E((\xi + \eta)^2) = E(\xi^2 + 2\xi\eta + \eta^2) = E(\xi^2) + 2E(\xi)E(\eta) + E(\eta^2),$$

$$(E(\xi + \eta))^2 = (E(\xi) + E(\eta))^2 = E(\xi)^2 + 2E(\xi)E(\eta) + E(\eta)^2.$$

Therefore

$$E((\xi + \eta)^2) - (E(\xi + \eta))^2 = E(\xi^2) - E(\xi)^2 + E(\eta^2) - E(\eta)^2 = D(\xi) + D(\eta).$$

In the formulation of Theorem 5.4 we didn't mention the existence of the expectation $E(\xi)$. It follows by the existence of $E(\xi^2)$. Namely

$$0 \le (|\xi| - 1)^2 = \xi^2 - 2|\xi| + 1,$$

$$|\xi| \le \frac{1}{2}(\xi^2 + 1).$$

Since $\frac{1}{2}(\xi^2 + 1)$ is integrable, also $|\xi|$ is integrable.

1.6. INDEPENDENT MEASUREMENTS

Let us measure some quantity. We realize a few measurements and count the arithmetic average of the measurements. A mathematical model of the experiments can be realized in finite sequence of independent random variables with the same probability distribution.

Theorem 1.6.1 *Let $\xi_1, \ldots, \xi_n$ be random variables with the same mean value $a = E(\xi_i)$, $i = 1, 2, \ldots, n$. Let $\bar{\xi} = \frac{1}{n}\sum_{i=1}^{n} \xi_i$. Then*

$$E(\bar{\xi}) = a.$$

Proof. We have

$$E(\bar{\xi}) = E\left(\frac{1}{n}\sum_{i=1}^{n} \xi_i\right) = \frac{1}{n}\sum_{i=1}^{n} E(\xi_i) = \frac{1}{n}\sum_{i=1}^{n} a = \frac{1}{n} \cdot na = a.$$

By a similar way the dispersion can be estimated.

Theorem 1.6.2 *Let $\xi_1, \ldots, \xi_n$ be independent random variables with the mean value $E(\xi_i) = a$, and the dispersion $D(\xi_i) = \sigma^2$ ($i = 1, 2, \ldots, n$). Then*

$$E\left(\frac{1}{n}\sum_{i=1}^{n} (\xi_i - a)^2\right) = \sigma^2.$$

Proof. We have

$$E\left(\frac{1}{n}\sum_{i=1}^{n}(\xi_i - a)^2\right) = \frac{1}{n}\sum_{i=1}^{n}E((\xi_i - a)^2) = \frac{1}{n}\sum_{i=1}^{n}D(\xi_i) = \frac{1}{n}\sum_{i=1}^{n}\sigma^2 = \sigma^2.$$

Of course, when we estimate σ^2 we don't know the number a. Therefore instead of a we use the arithmetic average

$$\bar{\xi} = \frac{1}{n}\sum_{i=1}^{n}\xi_i,$$

therefore we count

$$E\left(\sum_{i=1}^{n}(\xi_i - \bar{\xi})^2\right) = E\left(\sum_{i=1}^{n}(\xi_i^2 - 2\xi_i\bar{\xi} + \bar{\xi}^2)\right) = E\left(\sum_{i=1}^{n}\xi_i^2 - 2\bar{\xi}\sum_{i=1}^{n}\xi_i + n\bar{\xi}^2\right) =$$

$$= E\left((\sum_{i=1}^{n}\xi_i^2) - n\bar{\xi}^2\right) = \sum_{i=1}^{n}E(\xi_i^2) - nE(\bar{\xi}^2) =$$

$$= n(\sigma^2 + a^2) - nE(\bar{\xi}^2).$$

It remains to us to count

$$E(\bar{\xi}^2) = E\left(\left(\frac{1}{n}\sum_{i=1}^{n}\xi_i\right)^2\right) = \frac{1}{n^2}E\left(\sum_{i=1}^{n}\xi_i^2 + \sum_{i\neq j}\xi_i\xi_j\right) =$$

$$= \frac{1}{n^2}\left(\sum_{i=1}^{n}E(\xi_i^2) + \sum_{i\neq j}E(\xi_i)E(\xi_j)\right) =$$

$$= \frac{1}{n^2}n(\sigma^2 + a^2) + \frac{1}{n^2}n(n-1)a^2 = \frac{\sigma^2}{n} + \frac{a^2}{n} + a^2 - \frac{a^2}{n} =$$

$$= \frac{\sigma^2}{n} + a^2.$$

Therefore

$$E\left(\sum_{i=1}^{n}(\xi_i - \bar{\xi})^2\right) = n\sigma^2 + na^2 - n\frac{\sigma^2}{n} - na^2 = n\sigma^2 - \sigma^2 = \sigma^2(n-1).$$

Hence we have proved the following theorem.

Theorem 1.6.3 *Let* $\xi_1,\ldots,\xi_n$ *are independent random variables with the dispersion* $D(\xi_i) = \sigma^2$ $(i = 1,2,\ldots,n)$. *Then*

$$E\left(\frac{1}{n-1}\sum_{i=1}^{n}(\xi_i - \bar{\xi})^2\right) = \sigma^2.$$

1.7. INTERVAL ESTIMATION

We have seen that the arithmetical average $\bar{\xi}$ of measured values is a good estimation of mean value $E(\xi_i) = a$. Of course, measured values are concentrated about $E(\xi_i)$ some more nearlierly some least nearlierly. A theoretical model is Theorem 7.5. The mentioned law has been described by so called Gauss curve traditionally described in the formula

$$y = e^{-\frac{x^2}{2}}.$$

It is not possible to find to the density the primitive function by the help of elementary functions, of course, it is known that

$$\int_{-\infty}^{\infty} e^{-\frac{x^2}{2}}dx = \sqrt{2\pi}.$$

Definition 1.7.1 *A random variable ξ has the normal distribution with parameters $a \in \mathbb{R}$, $\sigma \in (0,\infty)(\xi\sim N(a,\sigma))$, if it is has the density*

$$f(x) = \frac{1}{\sigma\sqrt{2\pi}}e^{-\frac{(x-a)^2}{2\sigma^2}}.$$

If we denote

$$\varphi(t) = \frac{1}{\sqrt{2\pi}}e^{-\frac{t^2}{2}},$$

then

$$f(x) = \frac{1}{\sigma}\varphi\left(\frac{t-a}{\sigma}\right).$$

Theorem 1.7.2 *If ξ has the normal distribution with parameters a,*

$$\sigma(\xi\sim N(a,\sigma)), \text{ then}$$

$$E(\xi) = a, D(\xi) = \sigma^2.$$

Proof. The idea of the proof will be demonstrated in the case $a = 0, \sigma = 1$. In the general case, the proof is analogous. So

$$E(\xi) = \int_{-\infty}^{\infty} x \frac{1}{\sqrt{2\pi}} e^{-\frac{x^2}{2}} dx.$$

Since the function $y = xe^{-\frac{x^2}{2}}$ is add and integrable, we have

$$E(\xi) = \int_{-\infty}^{\infty} xe^{-\frac{x^2}{2}} dx = 0.$$

By the per partes methods we obtain

$$\int_{-\infty}^{\infty} x^2 e^{-\frac{x^2}{2}} dx = \int_{-\infty}^{\infty} x \left(xe^{-\frac{x^2}{2}} \right) dx = \left[x \cdot \left(-e^{-\frac{x^2}{2}} \right) \right]_{-\infty}^{\infty} - \int_{-\infty}^{\infty} - e^{-\frac{x^2}{2}} dx = 0 + \sqrt{2\pi},$$

hence

$$D(\xi) = \int_{-\infty}^{\infty} (x - 0)^2 \frac{e^{-\frac{x^2}{2}}}{\sqrt{2\pi}} dx = 1.$$

From the practical point of view it is usual to work with the distribution $N(0,1)$. Therefore the usual arithmetical average will be normed. Namely we know that

$$E(\bar{\xi}) = a, D(\bar{\xi}) = \sigma^2,$$

of course, under assumption that $\xi_1, \ldots, \xi_n$ are independent and integrable with the quadrat,

$$E(\xi_i) = a, D(\xi_i) = \sigma^2 \ (i = 1, 2, \ldots, n).$$

Lemma 1.7.3 If $E(\xi) = a, D(\xi) = \sigma^2, \sigma > 0$, and $\eta = \frac{1}{\sigma}(\xi - a)$, then $E(\eta) = 0$, $D(\eta) = 1$.

Proof. We have

$$E(\eta) = \frac{1}{\sigma} E(\xi - a) = \frac{1}{\sigma}(E(\xi) - a) = \frac{1}{\sigma}(a - a) = 0,$$

$$D(\eta) = E\left(\frac{1}{\sigma^2}(\xi - a)^2 \right) = \frac{1}{\sigma^2} E((\xi - a)^2) = \frac{1}{\sigma^2} \cdot \sigma^2 = 1.$$

Corollary 1.7.4 *Let $\xi_1, \ldots, \xi_n$ be independent random variables with the same distribution,* $E(\xi_i) = a, D(\xi_i) = \sigma^2, \sigma > 0 \ (i = 1, 2, \ldots, n)$. *Then*

$$E\left(\frac{1}{\sigma\sqrt{n}}\left(\sum_{i=1}^{n}\xi_i - na\right)\right) = E\left(\frac{\bar{\xi}-a}{\frac{\sigma}{\sqrt{n}}}\right) = 0,$$

$$D\left(\frac{1}{\sigma\sqrt{n}}\left(\sum_{i=1}^{n}\xi_i - na\right)\right) = D\left(\frac{\bar{\xi}-a}{\frac{\sigma}{\sqrt{n}}}\right) = 1.$$

The proposal theorem has the following form (one of the possibilities).

Theorem 1.7.5 (Central Limit Theorem) *Let $(\xi_n)_n$ be a sequence of independent equally distributed random variables with the mean value $E(\xi_n) = a$, and the dispersion $D(\xi_n) = \sigma^2$, $\sigma > 0 \ (i = 1, 2, \ldots)$. Then for any $x \in \mathbb{R}$ we have*

$$\lim_{n \to \infty} P\left(\left\{\omega; \frac{\sum_{i=1}^{n}\xi_i(\omega) - na}{\sigma\sqrt{n}} < x\right\}\right) = \frac{1}{\sqrt{2\pi}}\int_{-\infty}^{x} e^{-\frac{t^2}{2}} dt.$$

The proof ot the theorem will be discussed in the section 2.8.

Example 1.7.6 *Consider n independent measurements $\xi_1, \ldots, \xi_n$, and a real number $\delta > 0$. What is the probability of the fact that $|\bar{\xi} - a| < \delta$? We have*

$$P(|\bar{\xi} - a| < \delta) = P(\bar{\xi} - a < \delta) - P(\bar{\xi} - a \le -\delta) =$$

$$= P(\bar{\xi} - a < \tfrac{\varepsilon\sigma}{\sqrt{n}}) - P(\bar{\xi} - a \le -\tfrac{\varepsilon\sigma}{\sqrt{n}}),$$

where $\frac{\varepsilon\sigma}{\sqrt{n}} = \delta$, *hence* $\varepsilon = \frac{\sqrt{n}\delta}{\sigma}$. *Therefore*

$$P(|\bar{\xi} - a| < \delta) \doteq \int_{-\infty}^{\varepsilon} \frac{1}{\sqrt{2\pi}} e^{-\frac{t^2}{2}} dt - \int_{-\infty}^{-\varepsilon} \frac{1}{\sqrt{2\pi}} e^{-\frac{t^2}{2}} dt = \int_{-\varepsilon}^{\varepsilon} \varphi(t) dt = 2\int_{0}^{\varepsilon} \varphi(t) dt.$$

If we denote

$$\phi(\varepsilon) = \int_{0}^{\varepsilon} \varphi(t) dt,$$

then

$$P(|\bar{\xi} - a| < \delta) \doteq 2\phi(\varepsilon) = 2\phi\left(\frac{\sqrt{n}\delta}{\sigma}\right) = \alpha.$$

From Theorem 7.5 the classical Moivre and Laplace result follows. We have a random variable with the binomical distribution k_n with the values $0, 1, \ldots, n$ with the probabilities

$$p_k = P(\{\omega; \ k_n(\omega) = k\}) = n_{\ k}\ p^k (1 - p)^{n-k}.$$

Theorem 1.7.7 *Let k_n be a random variable with binomial distribution with the parameters n, p. Then for any $x \in \mathbb{R}$ there follows take equality*

$$\lim_{n \to \infty} P\left(\left\{\omega; \ \frac{k_n(\omega) - np}{\sqrt{np(1-p)}} < x\right\}\right) = \frac{1}{\sqrt{2\pi}} \int_{-\infty}^{x} e^{-\frac{t^2}{2}} dt.$$

Proof. Put $\xi_n = \chi_{A_n}$, where A_n are independent, and $P(A_n) = p$ $(n = 1, 2, \ldots)$. Then

$$E(\xi_n) = 1 \cdot +0 \cdot (1 - p) = p$$

$$D(\xi_n) = (1 - p)^2 \cdot p + (0 - p)^2 \cdot (1 - p) = p(1 - p)[1 - p + p] = p(1 - p),$$

hence $\sigma = \sqrt{np(1-p)}$. Moreover

$$\sum_{i=1}^{n} \xi_i = k_n.$$

Therefore

$$\frac{k_n(\omega) - np}{\sqrt{np(1-p)}} = \frac{\sum_{i=1}^{n} \xi_i(\omega) - na}{\sigma\sqrt{n}}.$$

1.8. PROOF OF CENTRAL LIMIT THEOREM

One of the possible techniques of the proof are so called characteristic functions. It is based on the Taylor development:

$$f(x) \sim f(0) + \frac{f'(0)}{1!} x + \frac{f''(0)}{2!} x^2 + \ldots + \frac{f^n(0)}{n!} x^n + \ldots$$

There are well known equalities

$$e^x = 1 + \frac{x}{1!} + \frac{x^2}{2!} + \ldots + \frac{x^n}{n!} + \ldots,$$

$$\sin x = x - \frac{x^3}{3!} + \frac{x^5}{5!} - \ldots + (-1)^{n-1} \frac{x^{2n-1}}{(2n-1)!} + \ldots,$$

$$\cos x = 1 - \frac{x^2}{2!} + \frac{x^4}{4!} - \ldots + (-1)^n \frac{x^{2n}}{(2n)!} + \ldots.$$

Some of the formulas hold for complex numbers, hence

$$e^{ix} = 1 + \frac{ix}{1!} + \frac{(ix)^2}{2!} + \frac{(ix)^3}{3!} + \frac{(ix)^4}{4!} + \frac{(ix)^5}{5!} + \ldots =$$

$$= 1 - \frac{x^2}{2!} + \frac{x^4}{4!} - \ldots + i\left(x - \frac{x^3}{3!} + \frac{x^5}{5!} - \ldots\right) =$$

$$= \cos x + i \sin x.$$

Definition 1.8.1 *If ξ is a random variable, then its characteristic function is the mapping $\psi: \mathbb{R} \to C$ defined by the formula*

$$\psi(t) = E\left(e^{it\xi}\right) = E(\cos t\xi) + iE(\sin t\xi),$$

of course, if the expectations exist.

Example 1.8.2 *Let ξ have the normal distribution $N(0,1)$. Then $\psi(t) = e^{-\frac{t^2}{2}}$. By the Definition 8.1 and Theorem 4.7*

$$\psi(t) = E\left(e^{i\xi}\right) = \int_{-\infty}^{\infty} e^{itx} \varphi(x) dx = \int_{-\infty}^{\infty} e^{itx} \frac{1}{\sqrt{2\pi}} e^{-\frac{x^2}{2}} dx.$$

But

$$itx - \frac{x^2}{2} = -\frac{1}{2}(x^2 - 2itx) = -\frac{1}{2}[(x - it)^2 - (it)^2] = -\frac{1}{2}(x - it)^2 - \frac{t^2}{2}.$$

Therefore

$$\psi(t) = \frac{1}{\sqrt{2\pi}} \int_{-\infty}^{\infty} e^{-\frac{(x-it)^2}{2}} e^{-\frac{t^2}{2}} d(x) = e^{-\frac{t^2}{2}} \int_{-\infty}^{\infty} \varphi(x) d(x) = e^{-\frac{t^2}{2}}.$$

In the next considerations we shall use the next lemma.

Lemma 1.8.3 *If $\lim_{n\to\infty} z_n = z$, then*

$$\lim_{n\to\infty} \ln\left(1 + \frac{z_n}{n}\right)^n = e^z.$$

Proof. We shall use the Taylor theorem for the function $\ln(1 + x)$, hence $f'(x) = \frac{1}{1+x}$, $f''(x) = -(1 + x)^{-2}$. If we put $a = 0$, then

$$\ln(1 + x) = f(0) + \frac{f'(0)}{1!}x + \frac{f''(c)}{2!}x^2 = 0 + x - \frac{1}{2(1+c)^2}\,x^2,$$

where c leads between 0 and x. Therefore

$$\ln\left(1 + \frac{z_n}{n}\right) = \frac{z_n}{n} - \frac{z_n^2}{2n^2(1+c)^2}.$$

Take n such that $\frac{|z_n^2|}{n} < q < 1$. Then

$$n\ln\left(1 + \frac{z_n}{n}\right) = z_n - \frac{z_n^2}{2n(1+c)^2}.$$

If we denote $\mathcal{R}_n = -\frac{z_n^2}{2n(1+c)^2}$, then

$$|\mathcal{R}_n| \leq \frac{z_n^2}{2n}$$

hence $\lim_{n\to\infty} \mathcal{R}_n = 0$. Therefore

$$\lim_{n\to\infty} \ln\left(1 + \frac{z_n}{n}\right)^n = \lim_{n\to\infty} z_n = z,$$

and therefore

$$\lim_{n\to\infty} \left(1 + \frac{z_n}{n}\right)^n = e^z.$$

Theorem 1.8.4 *Let $(\xi_n)_{n=1}^{\infty}$ be a sequence of independent random variables with the normal distribution, $E(\xi_n) = a$, $D(\xi_n) = \sigma^2, \sigma > 0$. Let ψ_n be the characteristic function of*

$$\frac{\sum_{j=1}^{n} \xi_j - na}{\sigma\sqrt{n}} \quad (n = 1, 2, \ldots).$$

Then

$$\lim_{n\to\infty} \psi(t) = e^{-\frac{t^2}{2}}$$

for any $t \in \mathbb{R}$.

Proof. Put

$$\eta_j = \frac{\xi_j - a}{\sigma}, j = 1,2,\dots$$

Then all η_j have the normal characteristic function

$$g(t) = E\left(e^{it\eta_j}\right), j = 1,2,\dots\dots$$

Let ψ_n be the characteristic function of

$$\frac{\sum_{j=1}^n \xi_j - na}{\sigma\sqrt{n}} \quad (n = 1,2,\dots).$$

Since η_j are independent, we have

$$\psi_n(\sqrt{n}t) = E\left(e^{it\sum_{j=1}^n \eta_j}\right) = E\left(\prod_{j=1}^n e^{it\eta_j}\right) = \prod_{j=1}^n (g(t)) = g(t)^n,$$

hence

$$\psi_n(t) = g\left(\frac{t}{\sqrt{n}}\right)^n.$$

But, if F is the distribution function of η_j, then

$$g(t) = E\left(e^{it\eta_j}\right) = \int_{-\infty}^{\infty} e^{itx} dF(x),$$

$$g'(t) = i\int_{-\infty}^{\infty} xe^{itx} dF(x),$$

$$g''(t) = -\int_{-\infty}^{\infty} x^2 e^{itx} dF(x),$$

hence

$$g(0) = E(1) = 1,$$

$$g'(0) = iE(\eta_i) = i \frac{1}{\sigma} E(\xi - a) = 0,$$

$$g''(c) = -\int_{-\infty}^{\infty} x^2 e^{icx} dF(x).$$

By the Taylor formula

$$g(t) = g(0) + \frac{g'(0)}{1!} t + \frac{g''(c)}{2!} t^2 = 1 - \frac{t^2}{2!} + \frac{t^2}{2!} - \frac{t^2}{2!} \int_{-\infty}^{\infty} x^2 e^{icx} dF(x) = 1 - \frac{t^2}{2!} + \rho(t).$$

Here

$$\frac{\rho(t)}{t^2} = \frac{1}{2} - \frac{1}{2} \int_{-\infty}^{\infty} x^2 e^{icx} dF(x),$$

where

$$\lim_{c \to 0} \int_{-\infty}^{\infty} x^2 e^{icx} dF(x) = \int_{-\infty}^{\infty} x^2 dF(x) = E(\eta^2) = \frac{1}{\sigma^2} E(\xi^2) = 1.$$

Therefore

$$\psi_n(t) = g\left(\frac{t}{\sqrt{n}}\right)^n = \left(1 - \frac{t^2}{2n} + \rho\left(\frac{t}{\sqrt{n}}\right)\right)^n =$$

$$= \left(1 + \frac{-\frac{t^2}{2} + n\rho\left(\frac{t}{\sqrt{n}}\right)}{n}\right)^n = \left(1 + \frac{z_n}{n}\right)^n.$$

But

$$\lim_{n \to \infty} z_n = \lim_{n \to \infty} \left(-\frac{t^2}{2} + t^2 \frac{\rho\left(\frac{t}{\sqrt{n}}\right)}{\frac{t^2}{n}}\right) = -\frac{t^2}{2} + t^2 \lim_{n \to 0} \frac{\rho(u)}{u^2} = -\frac{t^2}{2}.$$

Therefore by Lemma 8.3

$$\lim_{n \to \infty} \psi_n = \lim_{n \to \infty} \left(1 + \frac{z_n}{n}\right)^n = e^{-\frac{t^2}{2}}.$$

Of course, as we shall prove in the next section, if F_n is corresponding distribution function then

$$\lim_{n\to\infty} F_n(x) = \lim_{n\to\infty} \int_{-\infty}^{\infty} \varphi(t)d(t),$$

where

$$\varphi(t) = \frac{1}{\sqrt{2\pi}} e^{-\frac{t^2}{2}}.$$

Therefore Theorem 8.4 implies Theorem 7.2.

1.9. DISTRIBUTION AND CHARACTERISTIC FUNCTIONS

The aim of this section is the proof of the next theorem.

Theorem 1.9.1 *Let* $(F_n)_{n=1}^{\infty}$ *be a sequence of distribution functions and* ψ_n *be the corresponding characteristic functions, i.e. for any* $t \in \mathbb{R}$

$$\psi_n(t) = \int_{-\infty}^{\infty} e^{itx} dF_n(x).$$

If for any $t \in \mathbb{R}$

$$\lim_{n\to\infty} \psi_n(t) = e^{-\frac{t^2}{2}},$$

then

$$\lim_{n\to\infty} F_n(x) = \int_{-\infty}^{x} \frac{1}{\sqrt{2\pi}} e^{-\frac{t^2}{2}} d(t)$$

for any $x \in \mathbb{R}$.

For the proof of the Theorem 9.1 we need the following three propositions.

Proposition 1.9.2 *Let* $(G_k)_k$ *be a sequence of distribution functions. Then there exists a sequence* $\left(G_{k_l}\right)_{l=1}^{\infty}$ *(denote* $G_{k_l} = H_l$*) selected from* $(G_k)_k$ *and a left-continuous non-decreasing function* $G: \mathbb{R} \to [0,1]$, *such that*

$$\lim_{l\to\infty} H_l(x) = G(x)$$

for any $x \in \mathbb{R}$.

Proof. Let $Q = (x_i)_{i=1}^{\infty}$ be the set of all rational numbers. Because $(G_k(x_1))_{k=1}^{\infty}$ is bounded, there exists a convergent sequence $\left(G_k^{(1)}(x_1)\right)_{k=1}^{\infty}$ selected from $(G_k(x_1))_{k=1}^{\infty}$, such that there exists

$$\lim_{k \to \infty} G_k^{(1)}(x_1) = g(x_1).$$

Similarly, from the sequence $\left(G_k^{(1)}\right)_{k=1}^{\infty}$ we choose a sequence $\left(G_k^{(2)}\right)_{k=1}^{\infty}$ such that there exists

$$\lim_{k \to \infty} G_k^{(2)}(x_2) = g(x_2).$$

In this way we get the sequence of sequences

$$G_1^{(1)}, G_2^{(1)}, G_3^{(1)}, \dots$$

$$G_1^{(2)}, G_2^{(2)}, G_3^{(2)}, \dots$$

$$G_1^{(3)}, G_2^{(3)}, G_3^{(3)}, \dots$$

$$\dots$$

where the following sequence is selected from the previous one and for every i there exists

$$\lim_{k \to \infty} G_k^{(i)} = g(x_i).$$

Then the sequence $\left(G_n^{(n)}\right)_{n=1}^{\infty}$ is selected from the sequence $(G_k)_{k=1}^{\infty}$ and

$$\lim_{n \to \infty} G_n^{(n)}(x_i) = g(x_i)$$

for every i. We put $H_k = G_k^{(k)}$, thus

$$\lim_{k \to \infty} H_k(x_i) = g(x_i), i = 1, 2, \dots$$

Finally, it will be sufficient to put

$$G(x) = \lim_{u \to x^-} \sup\{g(x_i);\ x_i \le u\}.$$

Proposition 1.9.3 *Let* $(H_k)_{k=1}^{\infty}$ *be a sequence of distribution functions and* ψ_k *be the corresponding characteristic functions, e.g.*

$$\psi_k(t) = \int_{-\infty}^{\infty} e^{itx} dH_k(x).$$

Let

$$\lim_{k \to \infty} \psi_k(t) = e^{-\frac{t^2}{2}}, t \in \mathbb{R},$$

$$\lim_{k \to \infty} H_k(x) = G(x), x \in \mathbb{R}.$$

Then G *is a distribution function and*

$$\int_{-\infty}^{\infty} e^{itx} dG(x) = e^{-\frac{t^2}{2}}.$$

Proof. We have

$$\int_0^u e^{-\frac{t^2}{2}} dt = \int_0^u \lim_{k \to \infty} \psi_k(t) dt = \int_0^u \lim_{k \to \infty} \int_{-\infty}^{\infty} e^{itx} dH_k(x) dt =$$

$$= \lim_{k \to \infty} \int_{-\infty}^{\infty} \left[\int_0^u e^{itx} dt\right] dH_k(x) = \lim_{k \to \infty} \int_{-\infty}^{\infty} \frac{e^{iux}-1}{ix} dH_k(x) =$$

$$= \int_{-\infty}^{\infty} \frac{e^{iux}-1}{ix} dG(x) = \int_{-\infty}^{\infty} \left[\int_0^u e^{itx} dt\right] dG(x) =$$

$$= \int_0^u \left[\int_{-\infty}^{\infty} e^{itx} dG(x)\right] dt.$$

Hence

$$e^{-\frac{t^2}{2}} = \int_{-\infty}^{\infty} e^{itx} dG(x), t \in \mathbb{R}.$$

If we put in this equality $t = 0$, we get

$$1 = \lambda_G(\mathbb{R}),$$

thus G is the distribution function whose characteristic function is

$$\psi(t) = e^{-\frac{t^2}{2}}.$$

Proposition 1.9.4 *Let F, G be such distribution functions, that*

$$\int_{-\infty}^{\infty} e^{itx}\, dF(x) = \int_{-\infty}^{\infty} e^{itx}\, dG(x).$$

Then for any $x \in \mathbb{R}$ we have

$$F(x) = G(x).$$

Proof. First we consider a function ψ in the form

$$\psi(x) = \sum_{j=1}^{n} c_j e^{ixt_j}.$$

Obviously that

$$\int_{-\infty}^{\infty} \psi(x)\, dF(x) = \int_{-\infty}^{\infty} \psi(x)\, dG(x).$$

We know that every continuous function f on $\mathbb{R}$ which is null function and outside the compact interval can be approximate by the functions of the previous type. Therefore

$$\int_{-\infty}^{\infty} f(x)\, dF(x) = \int_{-\infty}^{\infty} f(x)\, dG(x).$$

But by such these functions can also be approximated functions of the type $\chi_{[a,b)}$. Therefore

$$\lambda_F([a,b)) = \int_{-\infty}^{\infty} \chi_{[a,b)}\, dF(x) = \int_{-\infty}^{\infty} \chi_{[a,b)}\, dG(x) = \lambda_G([a,b)).$$

Finally

$$F(x) = \lim_{n\to\infty} \lambda_F([x-n, x)) = \lim_{n\to\infty} \lambda_G([x-n, x)) = G(x).$$

Proof. [Proof of Theorem 9.1] Let $(G_k)_{k=1}^{\infty}$ be any sequence that is selected from the sequence $(F_n)_{n=1}^{\infty}$. Then by the Proposition 9.2 there exists such a non-decreasing continuous function $G: \mathbb{R} \to [0,1]$ and a sequence $(H_l)_{l=1}^{\infty}$ selected from a sequence $(G_k)_{k=1}^{\infty}$, that

$$G(x) = \lim_{l\to\infty} H_l(x), x \in \mathbb{R}.$$

By the Proposition 9.3 the function G is a distribution function and

$$\int_{-\infty}^{\infty} e^{itx} dG(x) = e^{-\frac{t^2}{2}}.$$

Define

$$F(x) = \int_{-\infty}^{x} \frac{1}{\sqrt{2\pi}} e^{-\frac{t^2}{2}} d(t).$$

But also

$$\int_{-\infty}^{\infty} e^{itx} dF(x) = e^{-\frac{t^2}{2}}.$$

Because from the equality of characteristic functions it follows the equality of distribution functions (Proposition 9.4), we can see that from any sequence $\left(F_{n_i}\right)_{i=1}^{\infty}$ selected from $(F_n)_{n=1}^{\infty}$ there exists such a sequence $\left(F_{n_{i_k}}\right)_{k=1}^{\infty}$ selected from $\left(F_{n_i}\right)_{i=1}^{\infty}$, that

$$\lim_{k \to \infty} F_{n_{i_k}}(x) = F(x) = \int_{-\infty}^{x} \frac{1}{\sqrt{2\pi}} e^{-\frac{t^2}{2}} d(t).$$

It follows that

$$\lim_{n \to \infty} F_n(x) = F(x)$$

for any $x \in \mathbb{R}$.

1.10 LAW OF LARGE NUMBERS

Central limit theorem gives us a good instrument for the estimation of the mean by the help of the arithmetic mean. Of course, since we have in disposition the Kolmogorov apparatus, therefore we mention two other known formulations.

Theorem 1.10.1 (Chebyshev's inequality) *Let ξ be a random variable having dispersion $D(\xi)$. Then for every $\varepsilon > 0$ there holds*

$$P(\{\omega; |\xi(\omega) - E(\xi)| \geq \varepsilon\}) \leq \frac{D(\xi)}{\varepsilon^2}.$$

Proof. Put $\mathcal{B} = \{\omega;\ |\xi(\omega) - E(\xi)|) \geq \varepsilon\}$. Then

$$D(\xi) = \int_\Omega (\xi - E(\xi))^2 d(P) \geq \int_\mathcal{B} (\xi - E(\xi))^2 d(P) \geq \varepsilon^2 P(\mathcal{B}).$$

Theorem 1.10.2 (3σ principle) *Let ξ be a random variable with positive dispersion $D(\xi) = \sigma^2$, $\sigma > 0$. Then*

$$P(\{\omega;\ |\xi(\omega) - E(\xi)| \geq 3\sigma\}) \leq \frac{1}{9}.$$

Proof. In Theorem 9.1 put $\varepsilon = 3\sigma$.

Theorem 1.10.3 (Week Law of Large Numbers) *Let $(\xi_n)_{n=1}^\infty$ be a sequence of independent random variables with the same distribution and with the dispersion. Then for every $\varepsilon > 0$ there holds*

$$\lim_{n \to \infty} P\left(\left\{\omega;\ \left|\frac{1}{n}\sum_{i=1}^n \xi_i - E(\xi_1)\right| \geq \varepsilon\right\}\right) = 0.$$

Proof. In Theorem 9.1 put $\xi = \frac{1}{n}\sum_{i=1}^n \xi_i$. Then

$$E(\xi) = \frac{1}{n}\sum_{i=1}^n E(\xi_i) = \frac{1}{n} \cdot n \cdot E(\xi_1) = E(\xi_1),$$

$$D(\xi) = \frac{1}{n^2}\sum_{i=1}^n D(\xi_i) = \frac{1}{n^2} \cdot n \cdot D(\xi_1) = \frac{1}{n}D(\xi_1).$$

Therefore

$$P\left(\left\{\omega;\ \left|\frac{1}{n}\sum_{i=1}^n \xi_i - E(\xi_1)\right| \geq \varepsilon\right\}\right) \leq \frac{1}{\varepsilon^2}\frac{1}{n}D(\xi_1).$$

Theorem 1.10.4 (Bernoulli's Law of Large Numbers) *If k_n is a random variable with binomial distribution with parameters $n \in N$, $p \in [0,1]$, ($n = 1,2,\dots$), then for any $\varepsilon > 0$ there holds*

$$\lim_{n \to \infty} P\left(\left\{\omega;\ \left|\frac{k_n(\omega)}{n} - p\right| \geq \varepsilon\right\}\right) = 0.$$

Proof. In Theorem 10.3 put $\xi_i = \chi_{A_i}$, where $A_1, A_2, \dots$ are independent events with $P(A_i) = p$ ($i = 1,2,\dots$).

Theorem 1.10.5 (Strong Law of Large Numbers) *Let $(\xi_n)_{n=1}^{\infty}$ be a sequence of independent random variables with the same distribution and with the dispersion. Then*

$$\lim_{n \to \infty} \left(\left\{ \omega; \tfrac{1}{n} \sum_{i=1}^{n} \xi_i(\omega) \right\} \right) = E(\xi_i)$$

for almost everywhere $\omega \in \Omega$.

1.11. CONDITIONAL PROBABILITY

Let the space Ω consists of n elements, its subset A has m elements, hence $P(A) = \frac{m}{n}$. Between these m elements k elements are such that belong also to B, hence

$$P(B/A) = \frac{k}{m} = \frac{\frac{k}{n}}{\frac{m}{n}} = \frac{P(A \cap B)}{P(A)}.$$

Therefore

$$P(A \cap B) = P(A) \cdot P(B/A).$$

If A, B are independent, then $P(A \cap B) = P(A) \cdot P(B)$, probability of the intersection is the product of probabilities. Generally, the probability of B depends on the condition A. Assume, that the conditions A are different. They can be described by a decomposition Γ of the set Ω

$$\Gamma = \{A_1, A_2, \ldots, A_n\},$$

where $P(A_i) > 0$ $(i = 1, 2, \ldots, n)$, $A_i \cap A_j = \emptyset$ $(i \neq j)$, $\bigcup_{i=1}^{n} A_i = \Omega$. The decomposition Γ creates the σ-algebra $S_0 = \sigma(\Gamma)$. If A_i is realized then the conditional probability of B is $P(B/A_i)$. Therefore as

$$P(B/S_0)(\omega) = \sum_{i=1}^{n} P(B/A_i)\chi_{A_i}(\omega).$$

The function $P(B/S_0): \Omega \to [0,1]$ has the following two properties:

1. The function $P(B/S_0)$ is S_0-measurable.

2. For any $E \in S_0$ there holds $\int_E P(B/S_0)dP = P(B \cap E)$.

Really, if $E \in S_0$, then $E = \bigcup_{i \in \alpha} A_i$, hence

$$\int_E P(B/S_0)dP = \int_\Omega \chi_E \sum_{i=1}^n P(B/A_i)\chi_{A_i}dP = \int_\Omega \sum_{i=1}^n P(B/A_i)\chi_{E \cap A_i}dP =$$

$$= \sum_{i \in \alpha} \int_\Omega P(B/A_i)\chi_{E \cap A_i}dP = \sum_{i \in \alpha} P(B/A_i)P(E \cap A_i) =$$

$$= \sum_{i \in \alpha} P(B \cap A_i) = P(B \cap \bigcup_{i \in \alpha} A_i) =$$

$$= P(B \cap E).$$

Of course, by the mentioned properties the conditional probability can be defined with respect to any σ-algebra S_0.

Theorem 1.11.1 *Let* (Ω, S, P) *be a probability space,* $S_0 \subset S$ *be a σ-algebra,* $B \in S$*. Then there exists S_0-measurable function* $f : \Omega \to [0,1]$ *such that*

$$\int_E f dP = P(E \cap B)$$

for any $E \in S_0$.

Proof. Define two measures μ, ν on S_0:

$$\mu(E) = P(E),$$

$$\nu(E) = P(E \cap B).$$

Then ν is absolutely continuous with respect to μ, *i.e.* $\mu(E) = 0 \Rightarrow \nu(E) = 0$. Therefore by the Radom-Nikodym theorem there exists an S_0-measurable function f such that

$$P(E \cap B) = \nu(E) = \int_E f d\mu = \int_E f dP$$

for any $E \in S_0$.

Of course, the function f is not defined uniquely, only "almost" uniquely.

Definition 1.11.2 *Let* f, g *be measurable functions defined on the space* (Ω, S, P)*. The functions* f, g *are equal μ-almost everywhere, if*

$$\mu(\{\omega \in \Omega; \ f(\omega) \neq g(\omega)\}) = 0.$$

Theorem 1.11.3 *Let* $\int_E f d\mu = v(E) = \int_E g d\mu$ *for all* $E \in S_0$. *Then* $f = g$ μ-*almost everywhere.*

Proof. Put $h = f - g$. Then h is S_0-measurable, and

$$\int_E h d\mu = 0$$

for all $E \in S_0$. If h is non-negative, simple, then

$$h = \sum_{i \in \alpha} \alpha_i \chi_{A_i},$$

$A_i \in S_0$ $(i = 1,2,\ldots,n)$, $A_i \cap A_j = \emptyset$ $(i \neq j)$. Then

$$0 = \int_E h d\mu = \sum_{i=1}^n \int_\Omega \alpha_i \chi_{A_i \cap E} d\mu = \sum_{i=1}^n \alpha_i \mu(A_i \cap E),$$

hence $\alpha_i = 0$, or $\mu(A_i \cap E) = 0$. Therefore $h = 0$ μ-almost everywhere.

If h is non-negative, S_0-measurable, we take simple h_n such that $0 \leq h_n \nearrow h$. Because

$$0 \leq \int_E h_n d\mu \leq \int_E h d\mu = 0,$$

we see that h_n are 0 μ-almost everywhere, and therefore h is 0 μ-almost

everywhere, too.

Finally for an arbitrary S_0-measurable function f, the sets

$$E^+ = \{\omega; \ h(\omega) > 0\}, E^- = \{\omega; \ h(\omega) < 0\}$$

are S_0-measurable,

$$\int_{E^+} h d\mu = 0, \int_{E^-} h d\mu = 0,$$

hence $h = f - g$ is 0 μ-almost everywhere on E^+ and also on E^- and also on E. It means that $f = g$ μ-almost everywhere.

Definition 1. 11.4 *Let* (Ω, S, P) *be a probability space,* $S_0 \subset S$ *be a* σ-*algebra,* $B \in S$. *Then* $P(B/S_0): \Omega \to \mathbb{R}$ *is any* S_0-*measurable function such that*

$$\int_E P(B/S_0)dP = P(B \cap E)$$

for all $E \in S_0$.

Theorem 1.11.5 *Let (Ω, S, P) be a probability space, $S_0 \subset S$ be a σ-algebra. Then*

1. *P-almost everywhere $P(\emptyset/S_0) = 0, P(\Omega/S_0) = 1$.*

2. *P-almost everywhere $0 \leq P(A/S_0) \leq 1, A \in S_0$.*

3. *If $A_n \in S$ $(n = 1,2,\dots), A_i \cap A_j = \emptyset$ $(i \neq j)$, then P-almost everywhere*

$$P(\cup_{n=1}^\infty A_n/S_0) = \sum_{n=1}^\infty P(A_n/S_0).$$

Proof. The first property follows by the fact that

$$\int_E 1dP = P(E) = P(E \cap \Omega), \int_E 0dP = 0 = P(E \cap \emptyset),$$

hence P-almost everywhere

$$P(\Omega/S_0) = 1, P(\emptyset/S_0) = 0.$$

For to prove the second property take $E = \{\omega; P(A/S_0) < 0\}$. The inequality $P(E) > 0$ leads to

$$P(E \cap A) = \int_E P(A/S_0)dP < 0$$

what is impossible. Similarly the second inequality can be proved. Finally, let $A_n \in S$ $(n = 1,2,\dots), A_i \cap A_j = \emptyset$ $(i \neq j)$. Let $E \in S_0$. Then

$$P(A_n \cap E) = \int_E P(A_n/S_0)dP.$$

Therefore

$$P\big((\cup_{n=1}^\infty A_n) \cap E\big) = P\big(\cup_{n=1}^\infty (A_n \cap E)\big) = \sum_{n=1}^\infty P(A_n \cap E) =$$

$$= \sum_{n=1}^\infty \int_E P(A_n/S_0)dP = \int_E \big(\sum_{n=1}^\infty P(A_n/S_0)\big)dP.$$

1.12. CONDITIONAL EXPECTATION

Similarly as the conditional probability also conditional expectation (mean value) can be introduced. Similarly as before Definition 11.4 existential Theorem 11.1 and Theorem 11.3 over were presented, now we first present an existential theorem.

Theorem 1.12.1 *Let* (Ω, S, P) *be a probability space,* $S_0 \subset S$ *be a* σ-algebra, $\xi: \Omega \to \mathbb{R}$ *be an integrable random variable. Then there exists an* S_0-measurable *function* $f: \Omega \to \mathbb{R}$ *such that*

$$\int_E f dP = \int_E \xi dP$$

for all $E \in S_0$. *If g is other such that function, then* $f = g$ *P-almost everywhere.*

Proof. We define on S_0 three measures $\mu, \nu_1, \nu_2: S_0 \to [0,1]$,

$$\mu(E) = P(E), \nu_1(E) = \int_E \xi^+ dP, \nu_2(E) = \int_E \xi^- dP.$$

Then ν_1 is absolutely continuous with respect to μ, ν_2 is absolutely continuous with respect to μ. Hence there exist S_0-measurable functions $f_1, f_2: \Omega \to \mathbb{R}$ such that

$$\nu_1(E) = \int_E f_1 d\mu, \nu_2(E) = \int_E f_2 d\mu.$$

If we put $f = f_1 - f_2$, then we for any $E \in S_0$ obtain

$$\int_E f dP = \int_E f_1 dP - \int_E f_2 dP = \nu_1(E) - \nu_2(E) = \int_E \xi^+ dP - \int_E \xi^- dP = \int_E \xi dP.$$

If also $\int_E g dP = \int_E \xi dP$ for any $E \in S_0$, then $\int_E g dP = \int_E f dP$ for every $E \in S_0$, hence $f = g$ P-almost everywhere by Theorem 11.3.

Definition 1.12.2 *Let* (Ω, S, P) *be a probability space,* $S_0 \subset S$ *be a* σ-algebra, $\xi: \Omega \to \mathbb{R}$ *be an integrable random variable. Then the conditional expectation* $E(\xi/S_0): \Omega \to \mathbb{R}$ *is every* S_0-measurable function such that

$$\int_E E(\xi/S_0) dP = \int_E f dP$$

for any $E \in S_0$.

Theorem 1.12.3 *Let* $(\Omega, \mathcal{S}, P)$ *be a probability space,* $\mathcal{S}_0 \subset \mathcal{S}$ *be a* σ-*algebra,* ξ *be a non-negative integrable random variable. Then*

$$E(\xi/\mathcal{S}_0) \geq 0$$

P-almost everywhere.

Proof. Put $E = \{\omega; E(\xi/\mathcal{S}_0)(\omega) < 0\}$. Then inequality $P(E) > 0$ leads to the relation

$$0 \leq \int_E E(\xi/\mathcal{S}_0)dP < 0$$

what is a contradiction.

Theorem 1.12.4 *Let* $(\Omega, \mathcal{S}, P)$ *be a probability space,* $\mathcal{S}_0 \subset \mathcal{S}$ *be a* σ-*algebra,* ξ, η *be integrable random variables,* $\alpha, \beta \in \mathbb{R}$. *Then*

$$E(\alpha\xi + \beta\eta/\mathcal{S}_0) = \alpha E(\xi/\mathcal{S}_0) + \beta E(\eta/\mathcal{S}_0).$$

Proof. For any $E \in \mathcal{S}_0$ we have

$$\int_E E(\xi/\mathcal{S}_0)dP = \int_E \xi dP, \int_E E(\eta/\mathcal{S}_0)dP = \int_E \eta dP.$$

Therefore

$$\int_E (\alpha\xi + \beta\eta)dP = \alpha \int_E \xi dP + \beta \int_E \eta dP = \alpha \int_E E(\xi/\mathcal{S}_0)dP + \beta \int_E E(\eta/\mathcal{S}_0) =$$

$$= \int_E (\alpha E(\xi/\mathcal{S}_0) + \beta E(\eta/\mathcal{S}_0))dP.$$

Theorem 1.12.5 *Let* $(\Omega, \mathcal{S}, P)$ *be a probability space,* $\mathcal{S}_0 \subset \mathcal{S}$ *be a* σ-*algebra,* $(\xi_n)_{n-1}^{\infty}$ *be a sequence of integrable random variables* $\xi_n \searrow 0$. *Then*

$$E(\xi/\mathcal{S}_0) \searrow 0$$

P-almost everywhere.

Proof. For $E \in \mathcal{S}_0$ we have $\chi_E \xi_n \searrow 0$, hence

$$0 = \lim_{n \to \infty} \int_E \xi_n dP = \lim_{n \to \infty} \int_E E(\xi_n/\mathcal{S}_0)dP = \int_E \left(\lim_{n \to \infty} E(\xi_n/\mathcal{S}_0) \right) dP.$$

1.13. LIMIT LAWS FOR MAXIMA

In Sections 2.8 and 2.9 we have studied limit laws for the arithmetical means

$$\bar{\xi}_n = \frac{1}{n}\sum_{i=1}^{n}\xi_i,$$

where $(\xi_n)_{n=1}^{\infty}$ is a sequence of independent random variables with the some distribution. We have seen that $(\bar{\xi}_n)_{n=1}^{\infty}$ converges to the normal distribution. In the present section we shall study an analogous question for the maxima

$$\eta_n = max(\xi_1,\ldots,\xi_n).$$

While the arithmetical means converges to the normal distribution, the maxima converge to the another distributions of exponential type [28, 31].

Definition 1.13.1 *A random variable ξ defined on the probability space (Ω, S, P) has the Fréchet distribution with the parameter $\alpha > 0$, if its distribution function $\Phi_\alpha: \mathbb{R} \to [0,1]$ is given by the formula*

$$\Phi_\alpha(x) = \begin{cases} 0, & \text{if } x \leq 0 \\ \exp(-x^{-\alpha}), & \text{if } x > 0. \end{cases}$$

Definition 1.13.2 *A random variable ξ defined on the probability space (Ω, S, P) has the Weibull distribution with the parameter $\alpha > 0$, if its distribution function $\Psi_\alpha: \mathbb{R} \to [0,1]$ is given by the formula*

$$\Psi_\alpha(x) = \begin{cases} \exp\{-(-x)^\alpha\}, & \text{if } x \leq 0 \\ 1, & \text{if } x > 0. \end{cases}$$

Definition 1.13.3 *A random variable ξ defined on the probability space (Ω, S, P) has the Gumbel distribution if its distribution function $\Lambda: \mathbb{R} \to [0,1]$ is given by the formula*

$$\Lambda(x) = \exp(-e^{-x}).$$

The mentioned distributions can be characterized by some real functions: If H is any of the mentioned distribution functions, then for any $t > 0$ there exist real numbers $\gamma(t) > 0$, and $\delta(t) \in \mathbb{R}$ such that

$$H^t(x) = H(\gamma(t)x + \delta(t)). \tag{1.1}$$

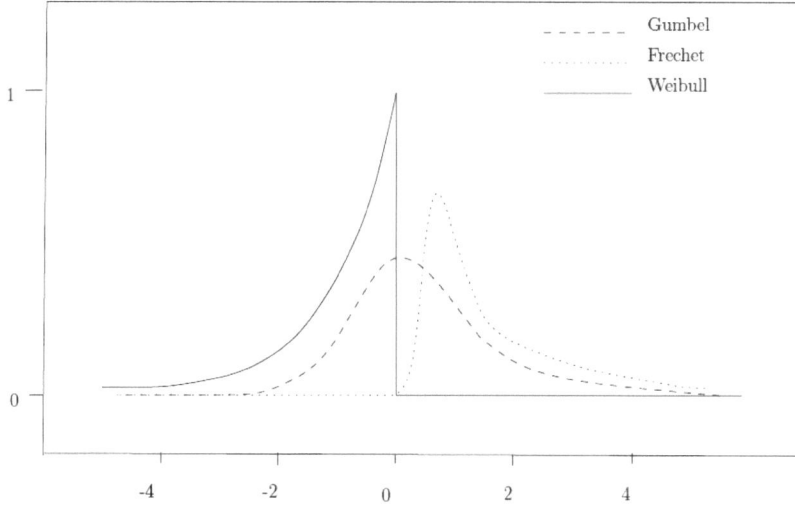

Source: The author of the figure is R. Bartková. The figure has not been published before in any other publication

Fig. (1). Densities of the standard extreme value distribution $\alpha = 1$.

Example 1.13.4 *Let Φ_α be the distribution function of Fréchet distribution. For $t > 0$ put*

$$\gamma(t) = t^{-\frac{1}{\alpha}}, \delta(t) = 0.$$

For $x \leq 0$

$$\Phi_\alpha^t(x) = 0 = \Phi_\alpha(\gamma(t)x + \delta(t)).$$

For $x > 0$ we have

$$\Phi_\alpha^t(x) = \exp(-tx^{-\alpha}).$$

On the other hand

$$\Phi_\alpha(\gamma(t)x + \delta(t)) = \Phi_\alpha\left(t^{-\frac{1}{\alpha}}x\right) = \exp\left(-\left(t^{-\frac{1}{\alpha}}x\right)^{-\alpha}\right) = \exp(-tx^{-\alpha}).$$

Example 1.13.5 *Let Ψ_α be the distribution function of Weibull distribution. For $t > 0$ put*

$$\gamma(t) = t^{\frac{1}{\alpha}}, \delta(t) = 0.$$

For $x \leq 0$

$$\Psi_\alpha^t(x) = \exp(-t(-x)^{-\alpha}),$$

and

$$\Psi_\alpha(\gamma(t) + \delta(t)) = \Psi_\alpha\left(t^{\frac{1}{\alpha}}x\right) = \exp\left(-\left(-t^{-\frac{1}{\alpha}}x\right)^\alpha\right) = \exp(-t(-x)^\alpha).$$

Of course, for $x > 0$, *we have also* $\gamma(t)x > 0$ *hence*

$$\Psi_\alpha^t(x) = 1 = \Psi_\alpha(\gamma(t) + \delta(t)).$$

Example 1.13.6 *Let* Λ *be the distribution function of Gumbel distribution. For* $t > 0$ *put*

$$\gamma(t) = 1, \delta(t) = \ln t.$$

Then

$$\Lambda^t(x) = \exp(-te^{-x}),$$

and

$$\Lambda(\gamma(t)x + \delta(t)) = \Lambda(x + \ln t) = \exp\left(-e^{(-x+\ln t)}\right) = \exp\left(-e^{\ln t}e^{-x}\right) = \exp(-te^{-x}).$$

Theorem 1.13.7 *Let* $\xi_1, \ldots, \xi_n$ *be independent random variables with the some distribution function* $F \colon \mathbb{R} \to [0,1]$. *Then the variable* $\eta_n = max(\xi_1, \ldots, \xi_n)$ *has the distribution function* $F^n \colon \mathbb{R} \to [0,1]$, *where*

$$F^n(x) = P(\{\omega; \eta_n(\omega) < x\}),$$

$x \in \mathbb{R}$.

Proof. We have

$$P(\{\omega; \eta_n(\omega) < x\}) = P(\{\omega; max(\xi_1, \ldots, \xi_n) < x\}) = P(\cap_{i=1}^n \{\omega; \xi_i(\omega) < x\}) =$$

$$= \prod_{i=1}^n P(\{\omega; \xi_i(\omega) < x\}) = \prod_{i=1}^n F(x) = F^n(x).$$

Theorem 1.13.8 *Let* $(\xi_n)_{n=1}^{\infty}$ *be a sequence of independent random variables with the same distribution function* $F \colon \mathbb{R} \to [0,1]$. *Put* $\eta_n = max(\xi_1, \ldots, \xi_n)$, $n = 1, 2, \ldots$ *Assume that there are* $a_n > 0, b_n$ *such that*

$$\lim_{n\to\infty} P\left(\left\{\omega; \; \frac{\eta_n(\omega) - b_n}{a_n} < x\right\}\right) = \lim_{n\to\infty} F^n(a_n x + b) = H(x),$$

where $H \colon \mathbb{R} \to [0,1]$ *is a continuous distribution function increasing on an interval. Then there are* $\gamma(t) > 0$, $\delta(t) \in \mathbb{R}$ *such that*

$$H^t(x) = H(\gamma(t)x + \delta(t)) \tag{1.2}$$

for any $t > 0$ *and* $x \in \mathbb{R}$.

Proof. For $t > 0$ let $[t]$ be the greatest integer less or equal to t. Denote

$$\alpha_n = a_{[nt]}, \beta_n = b_{[nt]}.$$

Then

$$\lim_{n\to\infty} F^{[nt]}(a_n x + b_n) = H(x),$$

$$\lim_{n\to\infty} F^{[nt]}(\alpha_n x + \beta_n) = \lim_{n\to\infty} (F^n(\alpha_n x + \beta_n))^{\frac{[nt]}{n}} = H^t(x).$$

Fix t and put $y = H(x)$, hence $x = H^{-1}(y)$. Then

$$\lim_{n\to\infty} (a_n x + b_n - F^{-n}(y)) = 0,$$

hence

$$\lim_{n\to\infty} \frac{F^{-n}(y) - b_n}{a_n} = H^{-1}(y).$$

Similarly

$$\lim_{n\to\infty} \frac{F^{-n}(y) - \beta_n}{\alpha_n} = H^{-t}(y).$$

If we choose two different values y_1, y_2 then there are

$$\lim_{n\to\infty} \frac{F^{-n}(y_1)-F^{-n}(y_2)}{a_n}, \; \lim_{n\to\infty} \frac{F^{-n}(y_1)-F^{-n}(y_2)}{\alpha_n}$$

and hence also there exists the limit of the division

$$\lim_{n\to\infty} \frac{a_n}{\alpha_n} = \gamma(t).$$

Similarly by the equality

$$\frac{\beta_n-b_n}{a_n} = \frac{F_n^{-1}(y)-b_n}{a_n} - \frac{F_n^{-1}(y)-\beta_n}{\alpha_n} \cdot \frac{\alpha_n}{a_n},$$

there follows the existence of

$$\lim_{n\to\infty} \frac{\beta_n-b_n}{a_n} = \delta(t).$$

Finally

$$H^t(x) = \lim_{n\to\infty} F^n(\alpha_n x + \beta_n) = \lim_{n\to\infty} F^n\left(\left(\frac{\alpha_n}{a_n}x + \frac{\beta_n-b_n}{a_n}\right)a_n + b_n\right) =$$

$$= H(\gamma(t)x + \delta(t)).$$

Theorem 1.13.9 *Let the assumptions of Theorem 13.8 be satisfied. Then for any* $t, s > 0$

$$\gamma(ts) = \gamma(t)\gamma(s),$$

$$\delta(ts) = \gamma(t)\delta(s) + \delta(t).$$

Proof. We have

$$H^{ts}(x) = H(\gamma(ts)x + \delta(ts)).$$

On the other hand

$$H^{ts}(x) = (H^s(x))^t = H^t(\gamma(s)x + \delta(s)) =$$

$$= H(\gamma(t)(\gamma(s)x + \delta(s)) + \delta(t)) =$$

$$= H(\gamma(t)\gamma(s)x + \gamma(t)\delta(s) + \delta(t)).$$

Therefore

$$\gamma(ts) = \gamma(t)\gamma(s) \tag{1.3}$$

$$\delta(ts) = \gamma(t)\delta(s) + \delta(t). \tag{1.4}$$

We have seen that it suffices to solve equalities 2.2, 2.3 and 2.4. First we shall look for the functional equation 2.3, so-called Hamel equation.

Theorem 1.13.10 *Let $\gamma: (0,\infty) \to (0,\infty)$ satisfy 2.3. Then there exists a such that*

$$\gamma(t) = t^a, t > 0.$$

Proof. Put

$$\varphi(t) = \ln\gamma(e^t).$$

Then

$$\varphi(u + v) = \ln\gamma(e^{t+v}) = \ln\gamma(e^u \cdot e^v) =$$

$$= \ln\gamma(e^u)\gamma(e^v) = \ln\gamma(e^u) + \ln\gamma(e^v) =$$

$$= \varphi(u) + \varphi(v),$$

hence φ satisfies the famous Cauchy equation. Therefore

$$\varphi(u) = u\varphi(1),$$

$$\ln\gamma(e^u) = u\varphi(1).$$

Put $\varphi(1) = a$, $e^u = t$, hence $u = \ln t$. Then

$$\gamma(t) = e^{u^a} = \left(e^{\ln t}\right)^a = t^a.$$

Now we are able to present and prove the main result of this section.

Theorem 1.13.11 *Let $(\xi_n)_{n=1}^\infty$ be a sequence of independent random variables with the same distribution function $F: \mathbb{R} \to [0,1]$. Put*

$$\eta_n = \max(\xi_1, \ldots, \xi_n), n = 1, 2, \ldots$$

Let there exist $a_n > 0, b_n \in \mathbb{R}$ such that

$$\lim_{n \to \infty} P\left(\left\{\omega; \frac{\eta_n(\omega) - b_n}{a_n} < x\right\}\right) = \lim_{n \to \infty} F^n(a_n x + b_n) = H(x),$$

where $H: \mathbb{R} \to [0,1]$ is a continuous distribution function, increasing on an interval. Then H has one of three distributions mentioned in Definitions 13.1-13.3 (Fréchet, Weibull, Gumbel).

Proof. We have obtained

$$H^t(x) = H(\gamma(t)x + \delta(t))$$

with 2.3 and 2.4. We know that

$$\gamma(t) = t^a.$$

We shall consider three cases with respect to a: $a = 0, a < 0, a > 0$. First let $a = 0$, *i.e.* $\gamma(t) = 1$. Then

$$\delta(ts) = \delta(t) + \delta(s).$$

Put

$$\psi(t) = e^{\delta(t)}.$$

Then

$$\psi(ts) = e^{\delta(ts)} = e^{(\delta(t) + \delta(s))} = e^{\delta(t)} \cdot e^{\delta(s)} = \psi(t) \cdot \psi(s).$$

Of course, by Theorem 13.10 there exists β such that

$$\psi(t) = t^\beta, t > 0.$$

Then

$$\delta(t) = \ln\psi(t) = \beta\ln t,$$

and 2.2 has the form

$$H^t(x) = H(x + \beta\ln t). \tag{1.5}$$

By 2.5 we see that H cannot have values 0 or 1, and that $0 < H(x) < 1$ for all $x \in \mathbb{R}$. Since $H(0) \in (0,1)$, there exist $c \in \mathbb{R}$ such that

$$H(0) = e^{-e^{-c}}.$$

Put $\beta \ln t = u$, *i.e.* $t = e^{\frac{u}{\beta}}$. Then

$$H^t(0) = e^{-e^{-c}t},$$

$$H(u) = H(\beta \ln t) = H^t(0) = e^{-e^{-c}e^{\frac{u}{\beta}}} = \exp\left(-e^{-(\beta^{-1}u+c)}\right) =$$

$$= \Lambda(\beta^{-1}u + c).$$

Now let $a < 0$. We have

$$\delta(ts) = \gamma(t)\delta(s) + \delta(t) = \gamma(s)\delta(t) + \delta(s),$$

hence

$$\delta(s)(1 - \gamma(t)) = \delta(t)(1 - \gamma(s))$$

or

$$\frac{\delta(s)}{1-\gamma(s)} = \frac{\delta(t)}{1-\gamma(t)}.$$

Denote the constant by d, hence

$$\delta(t) = d(1 - \gamma(t)).$$

The formula 2.5 has now the form

$$H^t(x) = H\left(t^a x + d(1 - t^a)\right),$$

or

$$H^t(x) = H(t^a(x - d) + d).$$

Since the equality holds for every x, we have also

$$H^t(x + d) = H(t^a x + d).$$

If we put $x = 0$, then

$$H^t(d) = H(d),$$

and $H(d)$ can be only 0 or 1. Of course $H(d) = 1$ is impossible, since the function

$$t \mapsto H^t(x + d)$$

would be decreasing but the function

$$t \mapsto H(t^u x + d)$$

would be increasing since $a < 0$. Therefore $H(d) = 0$ and

$$x \leq d \Rightarrow H(x) = 0.$$

On other hand

$$H(1 + d) \in (0,1),$$

hence we can put

$$H(1 + d) = e^{-p^\alpha},$$

where

$$\alpha = -\frac{1}{a} > 0.$$

Then

$$H(t^a + d) = H^t(1 + d) = e^{-tp^{-\alpha}}.$$

Denote $t^a + d = u$, hence

$$t = (u - d)^{\frac{1}{a}} = (u - d)^{-\alpha}.$$

Then

$$H(u) = e^{-(u-d)^{-\alpha} p^{-\alpha}} = \Phi_\alpha(p(u - d)).$$

Finally let $a > 0$. Then similarly as before we can obtain, that the function

$$t \mapsto \frac{\delta(t)}{1-\gamma(t)}$$

is constant. Let κ be the constant, hence

$$H^t(x + \kappa) = H(t^a + \kappa).$$

Now the equality

$$H^t(\kappa) = H(\kappa)$$

and $a > 0$ leads to the formula $H(\kappa) = 1$, hence

$$x \geq \kappa \Rightarrow H(x) = 1.$$

On the other hand $H(\kappa - 1) \in (0,1)$, hence we can choose

$$H(\kappa - 1) = e^{-p^{-\alpha}},$$

where $\alpha = \frac{1}{a} > 0$. Then

$$H(\kappa - t^a) = H^t(-1 + \kappa) = e^{-tp^\alpha},$$

or

$$H(u) = e^{-(\kappa+u)^{\frac{1}{a}} \cdot p^\alpha} = e^{-(\kappa+u)^\alpha p^\alpha} = \psi_\alpha(p(\kappa + u)).$$

1.14. PEAKS OVER THRESHOLD

As an alternative to studying the maxima of observation peaks over threshold method can be considered.

Consider the sequence $(\xi_n)_{n=1}^\infty$ of independent random variables with the some distribution $F: \mathbb{R} \to [0,1]$. Consider some high threshold, which we denote by w. For $w > 0$ we define conditional distribution function F_w in the following manner:

$$F_w(x) = P(\xi - w < x | \xi \geq w) =$$

$$= \frac{P(w \le \xi < x+w)}{P(\xi \ge w)} =$$

$$= \frac{F(x+w)-F(w)}{1-F(w)},$$

for $0 < x < \omega(F)$, where point $\omega(F) = \sup\{x; F(x) < 1\}$ is called right endpoint of distribution function F. Funkction F_w is called excess distribution.

Definition 1.14.1 *A random variable ξ defined on the probability space (Ω, S, P) has the Pareto distribution (GPD) with parameters α, β, if its distribution function $G_{\alpha,\beta}: \mathbb{R} \to [0,1]$ is given by the formula*

$$G_{\alpha,\beta}(\xi) = \begin{cases} 1 - \left(1 + \alpha \frac{\xi}{\beta}\right)^{-1/\alpha}, & \text{if } \alpha \ne 0, \\ 1 - \exp\left(-\frac{\xi}{\beta}\right), & \text{if } \alpha = 0, \end{cases}$$

where $\beta > 0$, $\alpha \ge 0$, if $x \in \langle 0, \infty)$ and $\alpha < 0$, if $x \in \langle 0, -\beta/\alpha\rangle$.

When parameter $\alpha > 0$, it takes the form of the ordinary Pareto distribution. If $\alpha = 0$, the GPD corresponds to exponential distribution, and Pareto II. type distribution for $\alpha < 0$.

Theorem 1.14.2 (Balkema, de Haan-Pickands) *For the sufficiently large w excess distribution F_w konverges to the GPD. Parameter $\beta = \beta(w)$ is depending on the threshold w, and for every $\alpha > 0$*

$$\lim_{w \to \omega(F)} \sup_{0 \le x \le \omega(F)-w} \left| F_w(x) - G_{\alpha,\beta_{(w)}}(x) \right| = 0.$$

Proof. For proof see Ref. [24].

CHAPTER 2

Probability Theory on Fuzzy Sets

Abstract: Similarly as the Kolmogorov probability theory in the first half of the 20th century, the Zadeh fuzzy set theory played a significant role in the second half of the 20th century. In this chapter we present probability theory on intuitionisic fuzzy sets as well as probability spaces on multivalued logic.

Keywords: Uncertainty, Fuzzy set, IF-set, IF-event, MV-algebras, Randomness, Observables.

Why does the Kolmogorov approach play such an important role in the probability theory and mathematical statistics? The reason is that Kolmogorov stated probability and statistics on a new and very effective formulation - set theory. For the first time in history basic notions of probability theory were defined precisely but simply. Thus, a random event got defined as a subset of a space, a random variable as a measurable function, and its mean value as an integral (abstract Lebesgue integral). More and more new results from the fuzzy generalization of the classical set theory are expected to be accomplished in the future.

2.1 FUZZY SETS

Any subset A of a given space Ω can be identified with its characteristic function

$$\chi_A : \Omega \to \{0,1\},$$

where

$$\chi_A(\omega) = 1,$$

if $\omega \in A$,

$$\chi_A(\omega) = 0,$$

if $\omega \notin A$. From the mathematical point of view a fuzzy set is a natural generalization of χ_A. It is a function [107].

$$\varphi_A : \Omega \to [0,1].$$

Evidently any set (*i.e.* two-valued function on Ω, $\chi_A : \Omega \to [0,1]$) is a special case of a fuzzy set (multi-valued function), $\varphi_A : \Omega \to [0,1]$.

Similarly as in the case of sets some binary operations can be introduced for fuzzy sets. They should correspond to the operations of union of sets, or intersections of sets. We shall present three possibilities here. They are extensions of the classical unions and intersections.

Definition 2.1.1 *Let $\mathcal{F}$ be set of all fuzzy sets defined on a space Ω, i.e.*

$$\mathcal{F} = \{f; f: \Omega \to [0,1]\}.$$

We define the following binary operations on $\mathcal{F}$: $\vee, \wedge, , \cdot, \oplus, \odot$:

$$f \vee g(\omega) = max\big(f(\omega), g(\omega)\big), f \wedge g(\omega) = min\big(f(\omega), g(\omega)\big),$$

$$fg(\omega) = f(\omega) + g(\omega) - f(\omega)g(\omega), f \cdot g(\omega) = f(\omega)g(\omega),$$

$$f \oplus g(\omega) = min(f(\omega) + g(\omega), 1), f \odot g(\omega) = max(f(\omega) + g(\omega) - 1,0).$$

Theorem 2.1.2 *If $f, g \in \mathcal{F}$, then $f \vee g \in \mathcal{F}$, $f \wedge g \in \mathcal{F}$, $fg \in \mathcal{F}$, $f \cdot g \in \mathcal{F}$, $f \oplus g \in \mathcal{F}$, $f \odot g \in \mathcal{F}$.*

Proof. Proof is straightforward. *E.g.*

$$f \vartriangle g(\omega) = f(\omega) + g(\omega)\big(1 - f(\omega)\big) \leq f(\omega) + 1 - f(\omega) = 1,$$

$$f \vartriangle g(\omega) = f(\omega) + g(\omega)\big(1 - f(\omega)\big) \geq f(\omega) \geq 0.$$

Another example

$$f \odot g(\omega) = max(f(\omega) + g(\omega) - 1,0) \leq max(f(\omega) + 1 - 1,0) \leq f(\omega) \leq 1,$$

$$f \odot g(\omega) = max(f(\omega) + g(\omega) - 1,0) \geq 0.$$

More interesting is the interpretation of the operations by the help of set operations.

Theorem 2.1.3 *If $f, g \in \mathcal{F}$ are characteristic functions $f = \chi_A$, $g = \chi_B$, then*

$$f \vee g = fg = f \oplus g = \chi_{A \cup B},$$

$$f \wedge g = f \cdot g = f \odot g = \chi_{A \cap B}.$$

Proof. Proof is straightforward.

Of course, another characterization of the operations mentioned above is by the help of operations with assertions. The operations $\vee, \triangle, \oplus$ corresponds to disjunction of assertions; $\wedge, \cdot, \odot$ corresponds to conjunction of assertions.

All above mentioned operations are commutative and associative. Moreover the couples $\vee, \wedge$ satisfies the distributive law.

Theorem 2.1.4 *Let $f, g, h \in \mathcal{F}$. Then*

$$(f \vee g) \wedge h = (f \wedge h) \vee (g \wedge h),$$

$$(f \wedge g) \vee h = (f \vee h) \wedge (g \vee h).$$

Proof. It is based on the following implication:

$$f(\omega) \leq g(\omega) \Rightarrow f \vee h(\omega) \leq g \vee h(\omega), f \wedge h(\omega) \leq g \wedge h(\omega).$$

Simple examples show that the couples $\triangle, \cdot,$ and $\oplus, \odot$ don't satisfy the distributive law.

Example 2.1.5 *Put $f = g = 1$, $h = \frac{1}{2}$ (constant functions). Then*

$$(f \triangle g) \cdot h = 1 \cdot \frac{1}{2} = \frac{1}{2},$$

$$(f \cdot h) \triangle (g \cdot h) = \frac{1}{2} \triangle \frac{1}{2} = \frac{3}{4}.$$

Example 2.1.6 *Put $f = g = 1$, $h = \frac{1}{2}$. Then*

$$(f \oplus g) \odot h = 1 \odot \frac{1}{2} = \frac{1}{2},$$

$$(f \odot h) \oplus (geh) = \frac{1}{2} \oplus \frac{1}{2} = 1.$$

On the other hand the de Morgan ruls hold for any couple $\vee, \wedge$, or $\triangle, \cdot$, or $\oplus, \odot$ resp.:

$$(A \cup B) = A' \cap B'.$$

Definition 2.1.7 *If $f \in \mathcal{F}$, then we define $\bar{f} = 1_\Omega - f$.*

Theorem 2.1.8 For any $f, g \in \mathcal{F}$ there holds

$$\overline{(f \vee g)} = \bar{f} \wedge \bar{g}, \overline{(f \wedge g)} = \bar{f} \vee \bar{g},$$

$$\overline{(f \vartriangle g)} = \bar{f} \cdot \bar{g}, \overline{f \cdot g} = \bar{f} \vartriangle \bar{g},$$

$$\overline{(f \oplus g)} = \bar{f} \odot \bar{g}, \overline{(f \odot g)} = \bar{f} \oplus \bar{g}$$

Proof. We have

$$\overline{(f \vee g)} = 1_\Omega - (f \vee g) = (1_\Omega - f) \wedge (1_\Omega - g) = \bar{f} \wedge \bar{g},$$

$$\overline{(f \wedge g)} = 1_\Omega - (f \wedge g) = (1_\Omega - f) \vee (1_\Omega - g) = \bar{f} \vee \bar{g}.$$

Similarly

$$\bar{f} \cdot \bar{g} = (1_\Omega - f) \cdot (1_\Omega - g) = 1_\Omega - f - g + fg = 1_\Omega - (f \vartriangle g) = \overline{f \vartriangle g},$$

and therefore also

$$\overline{f \cdot g} = \overline{\bar{\bar{f}} \cdot \bar{\bar{g}}} = \overline{\overline{\bar{f} \vartriangle \bar{g}}} = \bar{f} \vartriangle \bar{g}.$$

Finally

$$\overline{(f \oplus g)} = 1_\Omega - (f + g) \wedge 1_\Omega = (1_\Omega - f - g) \vee (1_\Omega - 1_\Omega) = (1_\Omega - f + 1_\Omega - g - 1_\Omega) \vee 0 = \bar{f} \odot \bar{g},$$

$$\overline{(f \odot g)} = \overline{\bar{\bar{f}} \odot \bar{\bar{g}}} = \overline{\overline{\bar{f} \oplus \bar{g}}} = \bar{f} \oplus \bar{g}.$$

2.2 INTUITIONISTIC FUZZY SETS

Of course, the main idea of the section is in the Atanassov generalization [4] of the previous Zadeh's concept [105]. There have been some discussion about terminology, therefore we shall speak only about IF-sets [2, 3, 77, 90].

Definition 2.2.1 *Let Ω be a non-empty set. By an IF-set we shall understand a pair of functions*

$$A = (\mu_A, \nu_A): \Omega \to [0,1]$$

where

$$\mu_A + \nu_A \leq 1_\Omega.$$

We call μ_A the membership function, ν_A the non-membership function and define the ordering

$$A \leq B \Leftrightarrow \mu_A \leq \mu_B, \nu_A \geq \nu_B.$$

Example 2.2.2 *An IF-set A is a mapping $A: \Omega \to \Delta$, where Δ is the triangle*

$$\Delta = \{(u,v) \in \mathbb{R}^2; 0 \leq u, 0 \leq v, u + v \leq 1\}$$

Fuzzy set φ can be considered as an IF-set, where

$$\mu_A = \varphi, \nu_A = 1_\Omega - \varphi,$$

hence fuzzy set can be considered as a mapping $\varphi_A: \Omega \to D$ to the segment

$$D = \{(u,v) \in \mathbb{R}^2; 0 \leq u, 0 \leq v, u + v = 1\},$$

and the classical set as a mapping $\psi: \Omega \to D_0$ to the two-poset set

$$D_0 = \{(0,1),(1,0)\}.$$

Similarly as in the fuzzy case also for IF-sets some binary operations can be introduced. Of course, if the membership functions is *e.g.* $\mu_A \oplus \mu_B$, how to define the non-membership function? It is reasonable to use the following method. Since

$$\mu_A \oplus \mu_B = (\mu_A + \mu_B) \wedge 1_\Omega,$$

we count

$$1_\Omega - (1_\Omega - \nu_A) \oplus (1_\Omega - \nu_B) = 1_\Omega - (1_\Omega - \nu_A + 1_\Omega - \nu_B) \wedge 1_\Omega =$$

$$= (1_\Omega - 1_\Omega + \nu_A - 1_\Omega + \nu_B) \vee (1_\Omega - 1_\Omega) =$$

$$= (\nu_A + \nu_B - 1_\Omega) \vee 0$$

and define the non-membership function as

$$\nu_A \odot \nu_B.$$

Theorem 2.2.3 *Let A, B are IF-sets (we shall write $A, B \in \mathcal{F}$), $A = (\mu_A, \nu_A)$, $B = (\mu_B, \nu_B)$. Define*

$$A \vee B = (\mu_A \vee \mu_B, \nu_A \wedge \nu_B), A \wedge B = (\mu_A \wedge \mu_B, \nu_A \vee \nu_B),$$

$$A \vartriangle B = (\mu_A \vartriangle \mu_B, \nu_A \cdot \nu_B), A \cdot B = (\mu_A \cdot \mu_B, \nu_A \vartriangle \nu_B),$$

$$A \oplus B = (\mu_A \oplus \mu_B, \nu_A \odot \nu_B), A \odot B = (\mu_A \odot \mu_B, \nu_A \oplus \nu_B).$$

Then

$$A \vee B \in \mathcal{F}, A \vee B \in \mathcal{F}, A \vartriangle B \in \mathcal{F}, A \cdot B \in \mathcal{F}, A \oplus B \in \mathcal{F}, A \odot B \in \mathcal{F}.$$

Proof. It suffices a little patience. As an illustration we show that $AB \in \mathcal{F}$:

$$\mu_A \vartriangle \mu_B + \nu_A \cdot \nu_B = \mu_A + \mu_B - \mu_A \mu_B + \nu_A \nu_B =$$

$$= \mu_A + \mu_B(1_\Omega - \mu_A) + \nu_A \nu_B \leq$$

$$\leq \mu_A + \mu_B(1_\Omega - \mu_A) + (1_\Omega - \mu_A)\nu_B =$$

$$= \mu_A + (1_\Omega - \mu_A)(\mu_B + \nu_B) \leq$$

$$\leq \mu_A + (1_\Omega - \mu_A) \cdot 1_\Omega = 1_\Omega.$$

Theorem 2.2.4 Let $A, B, C \in \mathcal{F}$. Then

$$(A \vee B) \wedge C = (A \wedge C) \vee (B \wedge C),$$

$$(A \wedge B) \vee C = (A \vee C) \wedge (B \vee C).$$

Proof. We shall prove the first equality. We have

$$(A \vee B) \wedge C = (\mu_A \vee \mu_B, \nu_A \wedge \nu_B) \wedge (\mu_C, \nu_C) =$$

$$= \big((\mu_A \vee \mu_B) \wedge \mu_C, (\nu_A \wedge \nu_B) \vee \nu_C\big) =$$

$$= \big((\mu_A \wedge \mu_C) \vee (\mu_B \wedge \mu_C), (\nu_A \vee \nu_C) \wedge (\nu_B \vee \nu_C)\big).$$

On the other hand

$$(A \wedge B) \vee (B \wedge C) = (\mu_A \wedge \mu_C, \nu_A \vee \nu_C) \vee (\mu_B \wedge \mu_C, \nu_B \vee \nu_C) =$$

$$= \left((\mu_A \wedge \mu_C) \vee (\mu_B \wedge \mu_C), (\nu_A \vee \nu_C) \wedge (\nu_B \vee \nu_C) \right).$$

The situation is similar to the fuzzy case. The above is the alone distributive law. Of course, de Morgan rule holds for any mentioned binary operations.

Definition 2.2.5 *For any $A = (\mu_A, \nu_A) \in \mathcal{F}$ we define $\bar{A} = (1_\Omega - \mu_A, 1_\Omega - \nu_A)$.*

Theorem 2.2.6 *Let $A, B \in \mathcal{F}$. Then*

$$\overline{(A \vee B)} = \bar{A} \wedge \bar{B}, \overline{(A \wedge B)} = \bar{A} \vee \bar{B},$$

$$\overline{(A \triangle B)} = \bar{A} \cdot \bar{B}, \overline{A \cdot B} = \bar{A} \triangle \bar{B},$$

$$\overline{(A \oplus B)} = \bar{A} \odot \bar{B}, \overline{(A \odot B)} = \bar{A} \oplus \bar{B}$$

Proof. We have

$$\overline{(A \vee B)} = (1_\Omega - \mu_A \vee \nu_B, 1_\Omega - \nu_A \wedge \nu_B) =$$

$$= \left((1_\Omega - \mu_A) \wedge (1_\Omega - \mu_B), (1_\Omega - \nu_A) \vee (1_\Omega - \nu_B) \right).$$

On the other hand

$$\bar{A} \wedge \bar{B} = (1_\Omega - \mu_A, 1_\Omega - \nu_A) \wedge (1_\Omega - \mu_B, 1_\Omega - \nu_B) =$$

$$= \left((1_\Omega - \mu_A) \wedge (1_\Omega - \mu_B), (1_\Omega - \nu_A) \vee (1_\Omega - \nu_B) \right).$$

Now using the equality $\bar{\bar{A}} = A, \bar{\bar{B}} = B, \bar{\bar{C}} = C$, we obtain

$$\overline{A \wedge B} = \overline{(\bar{A}) \wedge (\bar{B})} = \overline{\bar{A} \vee \bar{B}} = \bar{A} \vee \bar{B}.$$

Futher

$$\overline{(A \triangle B)} = (1_\Omega - (\mu_A + \mu_B - \mu_A \mu_B), 1_\Omega - \nu_A \nu_B) =$$

$$= \left((1_\Omega - \mu_A)(1_\Omega - \mu_B), 1_\Omega - \nu_A + 1_\Omega - \nu_B - (1_\Omega - \nu_A)(1_\Omega - \nu_B) \right) =$$

$$= (1_\Omega - \mu_A, 1_\Omega - \nu_A) \cdot (1_\Omega - \mu_B, 1_\Omega - \nu_B) = \bar{A} \cdot \bar{B},$$

$$\overline{A \cdot B} = \overline{\bar{A} \cdot \bar{\bar{B}}} = \overline{\overline{\bar{A} \vartriangle \bar{B}}} = \bar{A} \vartriangle \bar{B}.$$

Finally

$$\overline{A \oplus B} = (1_\Omega - \mu_A \oplus \mu_B, 1_\Omega - \nu_A \odot \nu_B) =$$

$$= \left((1_\Omega - \mu_A) \odot (1_\Omega - \mu_B), (1_\Omega - \nu_A) \oplus (1_\Omega - \nu_B)\right) = \bar{A} \odot \bar{B},$$

$$\overline{A \odot B} = \overline{\bar{A} \odot \bar{\bar{B}}} = \overline{\overline{\bar{A} \oplus \bar{B}}} = \bar{A} \oplus \bar{B}.$$

Example 2.2.7 *Let (Ω, S) be a measurable space. Consider the family $\mathcal{M}$ of all IF-sets $A = (\mu_A, \nu_A)$ such that $\mu_A \colon \Omega \to [0,1]$, $\nu_A \colon \Omega \to [0,1]$ are S-measurable. Then $\mathcal{M}$ is closed under the operations $\vee, \wedge, \vartriangle, \cdot, \oplus, \odot$.*

2.3 PROBABILITY ON IF-EVENTS

Definition 2.3.1 *Let (Ω, S) be a measurable space, i.e. Ω be a non-empty set, S be a σ-algebra. An IF-event is an S-measurable IF-set, i.e. IF-set*

$$A = (\mu_A, \nu_A)$$

such that $\mu_A \colon \Omega \to [0,1]$, $\nu_A \colon \Omega \to [0,1]$ are S-measurable ($I \subset \mathbb{R}$ is interval $\Rightarrow \mu_A^{-1}(I) \in S, \nu_A^{-1}(I) \in S$). By $\mathcal{F}$ we shall mean the family of all IF-events.

Probability $P(A)$ of an event $A \in \mathcal{F}$ has been defined constructively in [25], and then descriptively in [78]. This theory has been further developed in [41, 52, 74, 95].

Definition 2.3.2 *Let $\mathcal{J}$ be the family of all compact intervals in the unit interval, $\mathcal{J} = \{[a, b]; a, b \in \mathbb{R}, 0 \le a \le b \le 1\}$. Probability is a mapping*

$P \colon \mathcal{F} \to \mathcal{J}$ *satisfying the following conditions:*

(i) $P(1_\Omega, 0_\Omega) = [1,1]$, $P(0_\Omega, 1_\Omega) = [0,0]$,

(ii) $A \odot B = (0_\Omega, 1_\Omega) \Rightarrow P(A \oplus B) = P(A) + P(B)$,

(iii) $A_n \nearrow A \Rightarrow P(A_n) \nearrow P(A)$.

for every $A, B, A_n \in \mathcal{F}$.

Remark 2.3.3 $A_n = \left(\mu_{A_n}, \nu_{A_n}\right) \nearrow A = (\mu_A, \nu_A)$ *means that* $\mu_{A_n} \nearrow \mu_A$ *and* $\nu_{A_n} \searrow \nu_A$.

On the other hand, if $P(A_n) = [a_n, b_n]$, $P(A) = [a, b]$, then $P(A_n) \nearrow P(A)$ means $a_n \nearrow a$, $b_n \nearrow b$.

Theorem 2.3.4 *Let* $P: \mathcal{F} \to \mathcal{J}$ *be a probability. Denote*

$$P(A) = \langle P^{\#}(A), P^{\#}(A) \rangle$$

Then $P^{\#}(A): \mathcal{F} \to [0,1]$, $P^{\#}(A): \mathcal{F} \to [0,1]$ *have the following properties:*

(i) $P^{\#}(1_\Omega, 0_\Omega) = 1$, $P^{\#}(1_\Omega, 0_\Omega) = 1$, $P^{\#}(0_\Omega, 1_\Omega) = 0$, $P^{\#}(0_\Omega, 1_\Omega) = 0$,

(ii) $A \odot B = (0_\Omega, 1_\Omega) \Rightarrow P^{\#}(A \oplus B) = P^{\#}(A) + P^{\#}(B), P^{\#}(A \oplus B) = P^{\#}(A) + P^{\#}(B),$

(iii) $A_n \nearrow A \Rightarrow P^{\#}(A_n) \nearrow P^{\#}(A), P^{\#}(A_n) \nearrow P^{\#}(A).$

for every $A, B, A_n \in \mathcal{F}$.

Proof. It is straightforward. *E.g.* let $A \odot B = (0_\Omega, 1_\Omega)$. Then

$$P(A \oplus B) = \langle P^{\#}(A \oplus B), P^{\#}(A \oplus B) \rangle,$$

$$P(A) = \left(P^{\#}(A), P^{\#}(A)\right), P(B) = \left(P^{\#}(B), P^{\#}(B)\right)$$

hence

$$P(A) + P(B) = \langle P^{\#}(A) + P^{\#}(B), P^{\#}(A) + P^{\#}(B) \rangle$$

and therefore

$$P^{\#}(A \oplus B) = P^{\#}(A) + P^{\#}(B), P^{\#}(A \oplus B) = P^{\#}(A) + P^{\#}(B).$$

It is reasonable to study functions $m: \mathcal{F} \to [0,1]$. With respect to the terminology in quantum structures we shall name them states.

Definition 2.3.5 *[87, 88] Let* $\mathcal{F}$ *be the family of IF-events. A mapping* $m: \mathcal{F} \to \langle 0,1 \rangle$ *will be called a state if the following properties are satisfied:*

(i) $m(1_\Omega, 0_\Omega) = 1$, $m(0_\Omega, 1_\Omega) = 0$,

(ii) $A \odot B = (0_\Omega, 1_\Omega) \Rightarrow m(A \oplus B) = m(A) + m(B)$,

(iii) $A_n \nearrow A \Rightarrow m(A_n) \nearrow m(A)$,

for any $A, A_n, B \in \mathcal{F}$.

One of the nicest result concerning states on families of IF-events is the Cignoli representation theorem.

Theorem 2.3.6 *[89, 91] For any state $m: \mathcal{F} \to \langle 0,1 \rangle$ there exist probability measures $P, Q: S \to [0,1]$ and $\alpha \in [0,1]$ such that*

$$m(\mu_A, \nu_A) = \int_\Omega \mu_A dP + \alpha \left(1 - \int_\Omega (\mu_A + \nu_A) dQ \right)$$

for any $A = (\mu_A, \nu_A) \in \mathcal{F}$.

Proof. The main instrument in the proof is the following corollary of Definition 3.5 *(ii)*:

$$(f,g) \in \mathcal{F} \Rightarrow m(f,g) = m(f, 1_\Omega - f) + m(0_\Omega, f + g) \tag{2.1}$$

Define the mapping $P: S \to [0,1]$ by the formula

$$P(A) = m(\chi_A, 1_\Omega - \chi_A).$$

Let $A, B \in S, A \cap B = \emptyset$. Then $\chi_A + \chi_B \le 1$, hence

$$(\chi_A, 1_\Omega - \chi_A) \odot (\chi_B, 1_\Omega - \chi_B) = (0, 1_\Omega).$$

Therefore

$$P(A) + P(B) = m(\chi_A, 1_\Omega - \chi_A) + m(\chi_B, 1_\Omega - \chi_B) =$$

$$= m\big((\chi_A, 1_\Omega - \chi_A) \oplus (\chi_B, 1_\Omega - \chi_B)\big) =$$

$$= m\big((\chi_A + \chi_B, 1_\Omega - \chi_A - \chi_B)\big) =$$

$$= m(\chi_{A \cup B}, 1_\Omega - \chi_{A \cup B}) = P(A \cup B).$$

Let $A_n \in S \ (n = 1,2,\dots), A_n \nearrow A$. Then

$$\chi_{A_n} \nearrow \chi_A, 1_\Omega - \chi_{A_n} \searrow 1_\Omega - \chi_A,$$

hence by Definition 2.3.5 (*iii*)

$$P(A_n) = m(\chi_{A_n}, 1_\Omega - \chi_{A_n}) \nearrow m(\chi_A, 1_\Omega - \chi_A) = P(A).$$

Evidently $P(\Omega) = m(\chi_\Omega, 1_\Omega - \chi_\Omega) = m((1_\Omega, 0_\Omega)) = 1$, hence

$P: \mathcal{S} \to [0,1]$ is a probability measure.

Now we prove two identities. First the implication

$$A_1, \dots, A_n \in \mathcal{S}, \alpha_1, \dots, \alpha_n \in [0,1], A_i \cap A_j = \emptyset (i \neq j) \tag{2.2}$$

$$\Rightarrow m\left(\sum_{i=1}^n \alpha_i \chi_{A_i}, 1_\Omega - \sum_{i=1}^n \alpha_i \chi_{A_i}\right) = \sum_{i=1}^n m\left(\alpha_i \chi_{A_i}, 1_\Omega - \alpha_i \chi_{A_i}\right).$$

It can be proved by induction. The second identity is the following:

$$0 \leq \alpha, \beta \leq 1 \Rightarrow m(\alpha\beta\chi_A, 1_\Omega - \alpha\beta\chi_A) = \alpha m(\beta\chi_A, 1_\Omega - \beta\chi_A). \tag{2.3}$$

First it can be proved by induction the equality

$$km\left(\frac{1}{k}\beta\chi_A, 1_\Omega - \frac{1}{k}\beta\chi_A\right) = m(\beta\chi_A, 1_\Omega - \beta\chi_A)$$

holding for every $k \in \mathbb{N}$. Therefore

$$m\left(\frac{1}{k}\beta\chi_A, 1_\Omega - \frac{1}{k}\beta\chi_A\right) = \frac{1}{k}m(\beta\chi_A, 1_\Omega - \beta\chi_A),$$

$$m\left(\frac{p}{k}\beta\chi_A, 1_\Omega - \frac{p}{k}\beta\chi_A\right) = \frac{p}{k}m(\beta\chi_A, 1_\Omega - \beta\chi_A),$$

hence (2.3) holds for every rational α. Let $\alpha \in \mathbb{R}, \alpha \in [0,1]$. Take $\alpha_n \in \mathbb{Q}, \alpha_n \nearrow \alpha$. Then

$$\alpha_n \chi_A \nearrow \alpha\chi_A, 1_\Omega - \alpha_n\chi_A \searrow 1_\Omega - \alpha\chi_A.$$

Therefore

$$m(\alpha\beta\chi_A, 1_\Omega - \alpha\beta\chi_A) = \lim_{n \to \infty} m(\alpha_n\beta\chi_A, 1_\Omega - \alpha_n\beta\chi_A) =$$

$$\lim_{n \to \infty} \alpha_n m(\beta\chi_A, 1_\Omega - \beta\chi_A) = \alpha m(\beta\chi_A, 1_\Omega - \beta\chi_A),$$

hence (2.3) has been proved. Particularly, if we give $\beta = 1$, then

$$m(\alpha\chi_A, 1_\Omega - \alpha\chi_A) = \alpha m(\chi_A, 1_\Omega - \chi_A).$$

Let $f: \Omega \to [0,1]$ be simple, S-measurable

$$f = \sum_{i=1}^n \alpha_i\chi_{A_i}, A_i \in S(i = 1, \dots, n), A_i \cap A_j = \emptyset (i \neq j).$$

Combining (2.2), (2.3) and the definition of P we obtain

$$m(f, 1_\Omega - f) = \sum_{i=1}^n m\big(\alpha_i\chi_{A_i}, 1_\Omega - \alpha_i\chi_{A_i}\big) =$$

$$= \sum_{i=1}^n \alpha_i m\big(\chi_{A_i}, 1_\Omega - \chi_{A_i}\big) =$$

$$= \sum_{i=1}^n \alpha_i P(A_i) = \int_\Omega f dP,$$

hence

$$m(f, 1_\Omega - f) = \int_\Omega f dP$$

for any $f: \Omega \to [0,1]$ simple. If $f: \Omega \to [0,1]$ is an arbitrary S-measurable function, then there exist a sequence $(f_n)_n$ of simple non negative measurable functions such that $f_n \nearrow f$. Evidently $1_\Omega - f_n \searrow 1_\Omega - f$. Therefore

$$m(f, 1_\Omega - f) = \lim_{n\to\infty} m(f_n, 1_\Omega - f_n) = \int_\Omega f dP.$$

We have obtained

$$m(f, 1_\Omega - f) = \int_\Omega f dP \tag{2.4}$$

for any S-measurable $f: \Omega \to [0,1]$.

Now take our attention to the second term $m(0_\Omega, f + g)$ in the right sight of the equality (2.1). First define $M: S \to [0,1]$ by the formula

$$M(A) = m(0_\Omega, 1 - \chi_A).$$

As before it is possible to prove that M is a measure. Of course

$$M(\Omega) = m(0_\Omega, 0_\Omega) = \alpha \in [0,1].$$

Define $Q: S \to [0,1]$ by the formula

$$m(0_\Omega, 1_\Omega - \chi_A) = \alpha Q(A).$$

As before it is possible to prove

$$m(0_\Omega, 1_\Omega - f) = \alpha \int_\Omega f dQ,$$

for any S-measurable $f: \Omega \to [0,1]$, or

$$m(0_\Omega, h) = \alpha \int_\Omega (1_\Omega - h) dQ, \tag{2.5}$$

for any S-measurable $h: \Omega \to [0,1]$. Combining (3.1),(3.4) and (3.5) we obtain

$$m(A) = m\big((\mu_A, \nu_A)\big) = m\big((\mu_A, 1_\Omega - \mu_A)\big) + m\big((0_\Omega, \mu_A + \nu_A)\big) =$$

$$= \int_\Omega \mu_A dP + \alpha \big(1_\Omega - \int_\Omega (\mu_A + \nu_A) dQ\big).$$

Of course, since any fuzzy set is a special case of IF-set we can obtain also a representation theorem for state on fuzzy sets.

Theorem 2.3.7 *[86] Let (Ω, S) be a measurable space, $\mathcal{G}$ the family of all S-measurable functions $f: \Omega \to [0,1]$. Let $p: \mathcal{G} \to [0,1]$ satisfy the following conditions:*

$$p(1_\Omega) = 1;$$

$$f, g \in \mathcal{G}, f, g \le 1_\Omega \Rightarrow p(f + g) = p(f) + p(g);$$

$$f_n \in \mathcal{G} \ (n = 1,2, \dots), f_n \nearrow f \Rightarrow p(f_n) \nearrow p(f).$$

Then there exists a probability measure $P: S \to [0,1]$ such that

$$p(f) = \int_\Omega f dP$$

for any $f \in \mathcal{G}$.

Proof. Let $\mathcal{F}$ be the family of all IF-events A on Ω, i.e. $A = (\mu_A, \nu_A)$, where $\mu_A, \nu_A: \Omega \to [0,1]$ are S-measurable and $\mu_A + \nu_A \le 1_\Omega$. Define

$$m: \mathcal{F} \to [0,1]$$

by the formula

$$m\big((\mu_A, \nu_A)\big) = p(\mu_A).$$

Evidently $m(1_\Omega, 0_\Omega) = p(1_\Omega) = 1, m(0_\Omega, 1_\Omega) = p(0_\Omega) = 0$. Let

$A, B \in \mathcal{F}, A \odot B = (0_\Omega, 1_\Omega)$. Then

$$\big((\mu_A + \mu_B - 1_\Omega) \vee 0_\Omega, (\nu_A + \nu_B) \wedge 1_\Omega\big) = (0_\Omega, 1_\Omega),$$

hence

$$\mu_A + \mu_B \leq 1_\Omega.$$

Therefore

$$p(\mu_A + \mu_B) = p(\mu_A) + p(\mu_B) = m(A) + m(B).$$

On the other hand

$$A \oplus B = \big((\mu_A + \mu_B) \wedge 1_\Omega, (\nu_A + \nu_B - 1_\Omega) \vee 0_\Omega\big),$$

hence

$$m(A \oplus B) = p(\mu_A + \mu_B),$$

and $m \colon \mathcal{F} \to [0,1]$ is additive. Similarly it can be proved that m is continuous, hence Theorem 2.3.6 is applicable. Then for any $f \in \mathcal{G}$

$$p(f) = m(f, 1_\Omega - f) =$$

$$= \int_\Omega f dP + \alpha\big(1 - \int_\Omega (f + 1_\Omega - f) dQ\big) = \int_\Omega f dP.$$

2.4 OBSERVABLES

In Chapter 2 we presented the Kolmogorov concept of probability:

probability = measure

random variable = measurable function

mean value (expectation) = integral.

In this moment we shall realize something similar for IF-events:

state = measure

obervable = measurable function

mean value = integral.

Let (Ω, S) be given a measurable space, where Ω is a non-empty set, and S is a σ-algebra of subset of Ω. Let us remember that in the Kolmogorov theory random variable is a mapping $\xi: \Omega \rightarrow \mathbb{R}$ such that

$$\xi^{-1}(I) \in S$$

for any interval $I \subset \mathbb{R}$. An equivalent characterization has been given by σ-algebra $\mathcal{B}(\mathbb{R}) = \sigma(\mathcal{J})$ what is the smallest σ-algebra containing the family $\mathcal{J}$ of all intervals.

Definition 2.4.1 *An observable is a mapping $x: \sigma(\mathcal{J}) \rightarrow \mathcal{F}$ satisfying the following conditions*

(i) $x(\mathbb{R}) = (1_\Omega, 0_\Omega)$, $x(\emptyset) = (0_\Omega, 1_\Omega)$,

(ii) if $A, B \in \sigma(\mathcal{J})$, $A \cap B = \emptyset \Rightarrow x(A) \odot x(B) = (0_\Omega, 1_\Omega)$, $x(A \cup B) = x(A) \oplus x(B)$,

(iii) if $A, A_1, A_2, ... \in \sigma(\mathcal{J})$, $A_n \nearrow A \Rightarrow x(A_n) \nearrow x(A)$.

Theorem 2.4.2 *If $x: \sigma(\mathcal{J}) \rightarrow \mathcal{F}$ be an observable, and $m: \mathcal{F} \rightarrow [0,1]$ be a state, then*

$$m_x = m \circ x: \sigma(\mathcal{J}) \rightarrow [0,1]$$

defined by

$$m_x(A) = m\big(x(A)\big),$$

is a probability measure.

Proof. First

$$m_x(\mathbb{R}) = m\big(x(\mathbb{R})\big) = m(1_\Omega, 0_\Omega) = 1.$$

If $A \cap B = \emptyset$, then $x(A) \odot x(B) = (0_\Omega, 1_\Omega)$, hence

$$m_x(A \cup B) = m\big(x(A \cup B)\big) = m\big(x(A) \oplus x(B)\big) =$$

$$= m\big(x(A)\big) + m\big(x(B)\big) = m_x(A) + m_x(B).$$

Finally, $A_n \nearrow A$ implies $x(A_n) \nearrow x(A)$, hence

$$m_x(A_n) = m\big(x(A_n)\big) \nearrow m\big(x(A)\big) = m_x(A).$$

Theorem 2.4.3 *Let* $x\colon \sigma(\mathcal{J}) \to \mathcal{F}$ *be an observable,* $m\colon \mathcal{F} \to [0,1]$ *be a state. Define* $F\colon \mathbb{R} \to \langle 0,1 \rangle$ *by the formula*

$$F(u) = m\left(x\big((-\infty, u)\big)\right).$$

Then F is non-decreasing, left continuous in any point $u \in \mathbb{R}$, and

$$\lim_{u \to \infty} F(u) = 1, \ \lim_{u \to -\infty} F(u) = 0.$$

Proof. If $u < v$, then

$$x\big((-\infty, v)\big) = x\big((-\infty, u)\big) \oplus x\big((u, v)\big) \ge x\big((-\infty, u)\big),$$

hence

$$F(v) = m\left(x\big((-\infty, v)\big)\right) \ge m\left(x\big((-\infty, u)\big)\right) = F(u),$$

F is non-decreasing. If $u_n \nearrow u$, then

$$x\big((-\infty, u_n)\big) \nearrow x\big((-\infty, u)\big),$$

hence

$$F(u_n) = m\left(x\big((-\infty, u_n)\big)\right) \nearrow m\left(x\big((-\infty, u)\big)\right) = F(u),$$

F is left continuous in any $u \in \mathbb{R}$. Similarly, $u_n \nearrow \infty$ implies

$$x\big((-\infty, u_n)\big) \nearrow x\big((-\infty, \infty)\big) = (1_\Omega, 0_\Omega).$$

Therefore

$$F(u_n) = m\left(x\big((-\infty, u_n)\big)\right) \nearrow m\left(x\big((-\infty, \infty)\big)\right) = m(1_\Omega, 0_\Omega) = 1,$$

for every $u_n \nearrow \infty$, hence $\lim_{n \to \infty} F(u) = 1$. Similarly we obtain

$$u_n \searrow -\infty \Rightarrow -u_n \nearrow \infty,$$

hence

$$m\left(x\big((-u_n, u_n)\big)\right) \nearrow m\big(x(\mathbb{R})\big) = 1.$$

Now

$$1 = \lim_{n \to \infty} F(-u_n) = \lim_{n \to \infty} m\left(x\big((-u_n, u_n)\big)\right) + \lim_{n \to \infty} F(u_n) = 1 + \lim_{n \to \infty} F(u_n),$$

hence

$$\lim_{n \to -\infty} F(u_n) = 0, \text{ for any } u_n \searrow -\infty.$$

Combining Theorems 2.4.2 and 2.4.3 we obtain the following theorem.

Theorem 2.4.4 *Let $x: \mathcal{B}(\mathbb{R}) \to \mathcal{F}$ be an observable, $F: \mathbb{R} \to \langle 0,1 \rangle$ its distribution function, i.e.*

$$F(u) = m\left(x\big((-\infty, u)\big)\right),$$

$u \in \mathbb{R}$. Then there exists exactly one probability measure $\lambda_F: \mathcal{B}(\mathbb{R}) \to [0,1]$ such that

$$\lambda_F(\langle a, b \rangle) = F(b) - F(a).$$

Proof. For the existence put

$$\lambda_F = m_x.$$

Then

$$\lambda_F(\langle a, b \rangle) = m_x\big((-\infty, b)\big) - m_x\big((-\infty, a)\big) = F(b) - F(a).$$

If $\mu\colon \mathcal{B}(\mathbb{R}) \to [0,1]$ an arbitrary measure such that $\mu(\langle a, b)) = F(b) - F(a)$, then

$$\mu(\langle a, b)) = m\left(x\big((-\infty, b)\big)\right) - m\left(x\big((-\infty, a)\big)\right) =$$

$$= m_x\big((-\infty, b)\big) - m_x\big((-\infty, a)\big) =$$

$$= m_x(\langle a, b)) = \lambda_F(\langle a, b)).$$

Sine μ, and λ_F concide on intervals, they concide on $\sigma(\mathcal{J}) = \mathcal{B}(\mathbb{R})$.

Let us remember the transformation theorem (Theorem 1.4.1) As a consequence of it we obtain the formulas

$$E(x) = \int_{-\infty}^{\infty} x\,dF(x), D(x) = \int_{-\infty}^{\infty} (x - E(x))^2 dF(x) = \int_{-\infty}^{\infty} x^2 dF(x) - E(x)^2.$$

By this formulas the moments can be defined also for observables.

Definition 2.4.5 *Let $x\colon \mathcal{B}(\mathbb{R}) \to \mathcal{F}$ be an observable, $m\colon \mathcal{F} \to \langle 0,1\rangle$ a state, F be the corresponding distribution function. Then we define the mean value*

$$E(x) = \int_{-\infty}^{\infty} x\,d\lambda_F(x),$$

if the integral exists and the dispersion

$$D(x) = \int_{-\infty}^{\infty} x^2 d\lambda_F(x) - E(x)^2,$$

if the integral exists.

Example 2.4.6 *Let $x\colon \mathcal{B}(\mathbb{R}) \to \mathcal{F}$ be discrete, i.e. there exist $u_1, \dots, u_n \in \mathbb{R}$, and $p_1, \dots, p_n \in [0,1]$ such that*

$$F(u) = \sum \{u_i p_i; u_i \le u\}.$$

Then

$$E(x) = \sum_{i=1}^{n} u_i p_i, D(x) = \sum_{i=1}^{n} u_i^2 p_i - E(x)^2.$$

Example 2.4.7 *Let x be continous, i.e. there exists non negative integrable function f that*

$$F(u) = \int_{-\infty}^{u} f(t)dt,$$

for any $u \in \mathbb{R}$. Then

$$E(x) = \int_{-\infty}^{\infty} tf(t)dt, D(x) = \int_{-\infty}^{\infty} t^2 f(t)dt - E(x)^2.$$

2.5 INDEPEDENCE

In the classical case the independence has been defined by the identity

$$P(\xi^{-1}(A) \cap \eta^{-1}(B)) = P(\xi^{-1}(A))P(\eta^{-1}(B)).$$

In this case we kept at disposition the map $T = (\xi, \eta): \Omega \rightarrow \mathbb{R}^2$. Of course in the general case when two observables $x, y: \mathcal{B}(\mathbb{R}) \rightarrow \mathcal{F}$ are given we must construct something corresponding to T: a morphism from $\mathcal{B}(\mathbb{R}^2)$ to $\mathcal{F}$.

Definition 2.5.1 *Let* $x_1, \ldots, x_n: \sigma(\mathcal{J}) \rightarrow \mathcal{F}$ *be observables. By the joint observables of* $x_1, \ldots, x_n$ *we consider a mapping* $h: \sigma(\mathcal{J}^n) \rightarrow \mathcal{F}$ *($\mathcal{J}^n$ being the set of all intervals of* $\mathbb{R}^n$*) satisfying the following conditions:*

(i) $h(\mathbb{R}^n) = (1_\Omega, 0_\Omega),$

(ii) if $A, B \in \mathcal{J}$ *and* $A \cap B = \emptyset$ *then* $h(A) \odot h(B) = (0_\Omega, 1_\Omega)$ *and* $h(A \cup B) = h(A) \oplus h(B),$

(iii) if $A, A_1, A_2, \ldots \in \mathcal{J}$; $A_n \nearrow A$ *then* $h(A_n) \nearrow h(A),$

(iv) $h(C_1 \times \ldots \times C_n) = x_1(C_1). \cdots . x_n(C_n),$ *for any* $C_1, \ldots, C_n \in \mathcal{J}.$

Theorem 2.5.2 *For any observables* $x_1, \ldots, x_n: \sigma(\mathcal{J}) \rightarrow \mathcal{F}$ *there exists their joint observables* $h: \sigma(\mathcal{J}^n) \rightarrow \mathcal{F}.$

Proof. We shall prove it for $n = 2$. Consider two observables $x, y: \sigma(\mathcal{J}) \rightarrow \mathcal{F}$. Since $x(A) \in \mathcal{F}$, we shall write

$$x(A) = \left(x^\#(A), 1_\Omega - x^\#(A)\right),$$

and similarly

$$y(A) = \left(y^\#(B), 1_\Omega - y^\#(B)\right).$$

By the definition of product of IF-events we obtain

$$x(C) \cdot y(D) = \left(x^{\#}(C) \cdot y^{\#}(D), 1_{\Omega} - \left(1_{\Omega} - x^{\#}(C) \right)\left(1_{\Omega} - y^{\#}(D) \right) \right).$$

Therefore we shall construct similarly

$$h(K) = \left(h^{\#}(K), 1_{\Omega} - h^{\#}(K) \right).$$

Fix $\omega \in \Omega$ and define $\mu, \nu : \sigma(\mathcal{J}) \to [0,1]$ by

$$\mu(A) = x^{\#}(A)(\omega), \nu(B) = y^{\#}(B)(\omega).$$

Let $\mu \times \nu : \sigma(\mathcal{J}) \times \sigma(\mathcal{J}) = \sigma(\mathcal{J}^2) \to [0,1]$ be the product of probability measures μ, ν. Put

$$h^{\#}(K)(\omega) = \mu \times \nu(K).$$

Then

$$h^{\#}(C \times D)(\omega) = \mu \times \nu(C \times D) =$$

$$= \mu(C) \cdot \nu(D) = x^{\#}(C)(\omega) \cdot y^{\#}(D)(\omega)$$

hence

$$h^{\#}(C \times D) = x^{\#}(C) \cdot y^{\#}(D).$$

Analogously

$$1_{\Omega} - h^{\#}(C \times D) = \left(1_{\Omega} - x^{\#}(C) \right) \cdot \left(1_{\Omega} - y^{\#}(D) \right)$$

if we define

$$h(A) = \left(h^{\#}(A), 1_{\Omega} - h^{\#}(A) \right), A \in \sigma(\mathcal{J}^2),$$

then

$$h(C \times D) = \left(x^{\#}(C) \cdot y^{\#}(D), 1_{\Omega} - \left(1_{\Omega} - x^{\#}(C) \right)\left(1_{\Omega} - y^{\#}(D) \right) \right) =$$

$$= x(C) \cdot y(D).$$

Now we are able to define independence of observables and to use it for the study of probability laws.

Definition 2.5.3 *Let* $m: \mathcal{F} \to [0,1]$ *be a state,* $(x_n)_{n=1}^{\infty}$ *be a sequence of observables,* $h_n: \sigma(J^n) \to \mathcal{F}$ *be the joint observable of* $x_1, \ldots, x_n$ $(n = 1,2,\ldots)$. *Then* $(x_n)_{n=1}^{\infty}$ *called independent, if*

$$m\big(h_n(C_1 \times \ldots \times C_n)\big) = m\big(x_1(C_1)\big). \cdots . m\big(x_n(C_n)\big),$$

for any $n \in \mathbb{N}$ and any $C_1, \ldots, C_n \in \sigma(J)$.

2.6 CENTRAL LIMIT THEOREM

Recall the classical central limit theorem (Theorem 1.7.5) holding for independent observables $(\xi_n)_{n=1}^{\infty}$

$$\lim_{n \to \infty} P\left(\left\{\omega; \frac{\sum_{i=1}^{n} \xi_i(\omega) - na}{\sigma\sqrt{n}} < x\right\}\right) = \frac{1}{\sqrt{2\pi}} \int_{-\infty}^{x} e^{-\frac{t^2}{2}} dt.$$

Of course first we must prove the sum of observables

$$\frac{\sum_{i=1}^{n} x_i - na}{\sigma\sqrt{n}} : \sigma(J) \to \mathcal{F}.$$

Lemma 2.6.1 *Let* $a \in \mathbb{R}$, $\sigma > 0$, $n \in \mathbb{N}$, $x_1, \ldots, x_n: \sigma(J) \to \mathcal{F}$ *be observables,* $h: \mathcal{B}(\mathbb{R}^n) \to \mathcal{F}$ *be the joint observable of* $x_1, \ldots, x_n$, $g: \mathbb{R}^n \to \mathbb{R}$ *be a Borel measurable function. Put* $y: \sigma(J) \to \mathcal{F}$ *by*

$$y(A) = h\big(g^{-1}(A)\big), A \in \sigma(J).$$

Then y *is an observable.*

Proof. Evidently $y(\mathbb{R}) = h(\mathbb{R}^n) = (1_\Omega, 0_\Omega)$. Let $A, B \in \sigma(J)$, $A \cap B = \emptyset$. Then $g^{-1}(A) \in \sigma(J^n)$, $g^{-1}(B) \in \sigma(J^n)$ and $g^{-1}(A) \cap g^{-1}(B) = g^{-1}(A \cap B) = g^{-1}(\emptyset) = \emptyset$.

Therefore

$$y(A) e y(B) = h\big(g^{-1}(A)\big) \odot h\big(g^{-1}(B)\big) = (0_\Omega, 1_\Omega),$$

and

$$y(A \cup B) = h\big(g^{-1}(A \cup B)\big) = h\big(g^{-1}(A) \cup g^{-1}(B)\big) =$$

$$= h\big(g^{-1}(A)\big) \oplus h\big(g^{-1}(B)\big) = y(A) \oplus y(B).$$

Finally let $A_n \in \sigma(\mathcal{J})$ $(n = 1,2,\dots)$, $A_n \nearrow A$. Then $g^{-1}(A_n) \nearrow g^{-1}(A)$,

and

$$y(A_n) = h\big(g^{-1}(A_n)\big) \nearrow h\big(g^{-1}(A)\big) = y(A).$$

Definition 2.6.2 *Let* $x_1,\dots,x_n\colon \sigma(\mathcal{J}) \to \mathcal{F}$ *be observables,* h_n *be their join observable,* $a \in \mathbb{R}$, $\sigma > 0$, $g\colon \mathbb{R}^n \to \mathbb{R}$ *be defined by*

$$g(u_1, \dots, u_n) = \frac{1}{\sigma\sqrt{n}}\left(\textstyle\sum_{i=1}^n u_i - na\right).$$

Then we define

$$\frac{1}{\sigma\sqrt{n}}\left(\textstyle\sum_{i=1}^n x_i - na\right) = h_n \circ g^{-1}.$$

Theorem 2.6.3 *Let* $(x_n)_{n=1}^\infty$ *be a sequence of square integrable, equally distributed, independent observables,* $E(x_n) = a, \sigma^2(x_n) = \sigma^2$ $(n = 1,2,\dots)$. *Then for every* $t \in \mathbb{R}$

$$\lim_{n\to\infty} m\left(\frac{\frac{1}{n}\sum_{i=1}^n x_i - a}{\frac{\sigma}{\sqrt{n}}}\big((-\infty, t)\big)\right) = \frac{1}{\sqrt{2\pi}}\int_{-\infty}^t e^{-\frac{u^2}{2}}\,du.$$

Proof. We want to use Theorem 7.5. We shall consider the space $\mathbb{R}^{\mathbb{N}}$ of all sequences $(t_n)_{n=1}^\infty$ of real numbers, and the family $\mathcal{C}$ of all cylinders in the space $\mathbb{R}^{\mathbb{N}}$. A set $A \subset \mathbb{R}^{\mathbb{N}}$ is a cylinder, if there exists $k \in \mathbb{N}$, and $B \in \sigma(\mathcal{J}^k)$ such that

$$A = \{(t_n)_{n=1}^\infty; (t_1, \dots, t_n) \in B\}.$$

Now $(\mathbb{R}^{\mathbb{N}}, \sigma(\mathcal{C}))$ is a measurable space and we shall construct a probability measure $P\colon \sigma(\mathcal{C}) \to [0,1]$ on the space.

First let $h_n\colon \sigma(\mathcal{J}^n) \to \mathcal{F}$ be the joint observable of $x_1, \dots, x_n$, and

$$\mu_n = m \circ h_n : \sigma(\mathcal{J}^n) \to [0,1].$$

Then μ_n is a probability measure for any n. Moreover

$$\mu_{n+1}|(\sigma(\mathcal{J}^n) \times \mathbb{R}) = \mu_n,$$

$n = 1,2, ...$ Then by the Kolmogorov consistency theorem there exists exactly one probability measure

$$P : \sigma(\mathcal{C}) \to [0,1]$$

such that

$$P(A) = \mu_k(B)$$

for any cylinder $A \in \mathcal{C}$ generated by $B \in \sigma(\mathcal{J}^k)$. So we obtained the probability space

$$(\mathbb{R}^{\mathbb{N}}, \sigma(\mathcal{C}), P).$$

Now we define $\xi_n : \mathbb{R}^{\mathbb{N}} \to \mathbb{R}$ be the equality

$$\xi_n((t_i)_{i=1}^{\infty}) = t_n$$

and $g_n : \mathbb{R}^n \to \mathbb{R}$ by the formula

$$g_n(u_1, ..., u_n) = \frac{1}{\sigma\sqrt{n}} (\sum_{i=1}^{n} u_i - na).$$

Further, let

$$\Pi_n : \mathbb{R}^{\mathbb{N}} \to \mathbb{R}^n$$

be defined by

$$\Pi_n((t_i)_{i=1}^{\infty}) = (t_1, ..., t_n)$$

$$\eta_n : \mathbb{R}^{\mathbb{N}} \to \mathbb{R}, \eta_n = g_n \circ \Pi_n$$

$$y_n : \sigma(\mathcal{J}^n) \to \mathcal{F}, y_n = h_n \circ g_n^{-1}.$$

Take $B \in \mathcal{B}(\mathbb{R})$. Then

$$m\big(y_n(B)\big) = m\left(h_n\big(g_n^{-1}(B)\big)\right) = P\left(\Pi_n^{-1}\big(g_n^{-1}(B)\big)\right) =$$

$$= P\big((g_n \circ \Pi_n)^{-1}(B)\big) = P\big(\eta_n^{-1}(B)\big).$$

Now for $C \in \sigma(\mathcal{J})$

$$m\big(x_n(C)\big) = m\big(h_n(\mathbb{R} \times \ldots \times \mathbb{R} \times C)\big) =$$

$$= P\big(\Pi_n^{-1}(\mathbb{R} \times \ldots \times \mathbb{R} \times C)\big) = P\big(\xi_n^{-1}(C)\big),$$

hence

$$E(\xi_n) = \int_{-\infty}^{\infty} t\, dP_{\xi_n}(t) = \int_{-\infty}^{\infty} t\, dm_{x_n}(t) = E(x_n) = a,$$

and

$$D(\xi_n) = D(x_n) = \sigma^2.$$

Moreover,

$$P\big(\xi_1^{-1}(C_1) \cap \ldots \cap \xi_n^{-1}(C_n)\big) = P\big(\Pi_n^{-1}(C_1 \times \ldots \times C_n)\big) =$$

$$= m\big(h_n(C_1 \times \ldots \times C_n)\big) =$$

$$= m\big(x_1(C_1)\big) \cdot \ldots \cdot m\big(x_n(C_n)\big) =$$

$$= P\big(\xi_1^{-1}(C_1)\big) \cdot \ldots \cdot P\big(\xi_n^{-1}(C_n)\big),$$

hence $\xi_1, \ldots, \xi_n$ are independent for every n. Moreover

$$m\left(\frac{\sqrt{n}}{\sigma}\big(\textstyle\sum_{i=1}^{n} x_i - a\big)\big((-\infty, t)\big)\right) = m\left(h_n\left(g_n^{-1}\big((-\infty, t)\big)\right)\right) =$$

$$= m\left(y_n\big((-\infty, t)\big)\right) =$$

$$= P\left(\eta_n^{-1}\big((-\infty, t)\big)\right) =$$

$$= P\left(\Big\{\omega; \frac{\sqrt{n}}{\sigma}\textstyle\sum_{i=1}^{n} \xi_i(\omega) - a < t\Big\}\right),$$

hence Theorem 1.7.5 can be applied.

2.7 EMBEDDING INTO MV-ALGEBRAS

MV-algebra is a new mathematical model for various problems of multi valued logic, where it plays a similar role as Boolean algebra for two valued logic.

Example 2.7.1 *Consider the unit interval* $[0,1]$ *with the usual operation* $\leq$, *two fixed elements* $0,1$ *and the Lukasiewicz operation* $\oplus, \odot$, *given by*

$$a \oplus b = min(a + b, 1), a \odot b = max(a + b - 1, 0).$$

By the Mundici theorem ([58]) MV-algebra can be characterized by the help of l-groups.

Definition 2.7.2 *An algebraic system* $(G, +, \leq)$ *is called l-group, if*

(i) $(G, +)$ *is and Abelian group,*

(ii) $(G, \leq)$ *is a lattice,*

(iii) $a \leq b \Rightarrow a + c \leq b + c$ *for any* a, b, c *in* G.

Consider the neutral element $0 \in G$ and a positive element $u > 0$. Then MV-algebra is an algebraic system $(G, \leq, 0, u, \oplus, \odot)$, where

$$a \oplus b = (a + b) \wedge u, a \odot b = (a + b - u) \vee 0.$$

In Example 7.1 there was $(G, +) = (\mathbb{R}, +)$, zero element was the neutral element, $\leq$ was the usual ordering, and $u = 1$. Now we shall embed the family $\mathcal{F}$ of IF-events to an MV-algebra.

Theorem 2.7.3 [75, 85] *Let* $(\Omega, \mathcal{S})$ *be a measurable space,* $\mathcal{F}$ *be the family of all IF-sets* $A = (\mu_A, \nu_A)$ *such that* μ_A, ν_A *are* $\mathcal{S}$-measurable. *Then there exists an MV-algebra* $\mathcal{M}$ *such that* $\mathcal{F} \subset \mathcal{M}$, *the operations* $\oplus, \odot$ *are extensions of operations on* $\mathcal{F}$ *and the ordering* $\leq$ *is an extension of the ordering in* $\mathcal{F}$.

Proof. Consider the group G of all mappings set $\mathbb{R}^2$ for the $A = (\mu_A, \nu_A)$ from Ω to $\mathbb{R}^2$. Of course the ordering $A \leq B$ must be defined as in $\mathcal{F}$, hence

$$G = \{A = (\mu_A, \nu_A); \mu_A: \Omega \to \mathbb{R}, \nu_A: \Omega \to \mathbb{R}, \mu_A, \nu_A \, \mathcal{S} - measurable\}$$

and

$$A \leq B \Leftrightarrow \mu_A \leq \mu_B, \nu_A \geq \nu_B.$$

The element $(0_\Omega, 1_\Omega)$ should be the neutral element. Therefore we define

$$A + B = \left(\mu_A + \mu_B, 1_\Omega - (1_\Omega - \nu_A + 1_\Omega - \nu_B)\right) = (\mu_A + \mu_B, \nu_A + \nu_B - 1_\Omega).$$

Actually $(G, +)$ is an Abelian group with $(0_\Omega, 1_\Omega)$ as the neutral element

$$A + (0_\Omega, 1_\Omega) = (\mu_A, \nu_A) + (0_\Omega, 1_\Omega) = (\mu_A + 0_\Omega, \nu_A + 1_\Omega - 1_\Omega) = (\mu_A, \nu_A).$$

Also $(G, \leq)$ is a lattice with

$$A \vee B = (\mu_A \vee \mu_B, \nu_A \wedge \nu_B),$$

$$A \wedge B = (\mu_A \wedge \mu_B, \nu_A \vee \nu_B).$$

Put $u = (1_\Omega, 0_\Omega)$ and define

$$M = \{A \in G; (0_\Omega, 1_\Omega) \leq A \leq (1_\Omega, 0_\Omega)\},$$

hence $0_\Omega \leq \mu_A \leq 1_\Omega, 0_\Omega \leq \nu_A \leq 1_\Omega$. It remains to us to compute the operations $\oplus$, $\odot$. Of course,

$$A \oplus B = (A + B) \wedge u =$$

$$= (\mu_A + \mu_B, \nu_A + \nu_B - 1_\Omega) \wedge (1_\Omega, 0_\Omega) =$$

$$= \left((\nu_A + \mu_B) \wedge 1_\Omega, (\nu_A + \nu_B - 1) \wedge 0_\Omega\right),$$

and

$$A \odot B = (A + B - u) \vee 0 =$$

$$= [(\mu_A + \mu_B, \nu_A + \nu_B - 1) - (1_\Omega, 0_\Omega)] \vee (0_\Omega, 1_\Omega) =$$

$$= (\mu_A + \mu_B - 1_\Omega, \nu_A + \nu_B - 1_\Omega - 0_\Omega + 1_\Omega) \vee (0_\Omega, 1_\Omega) =$$

$$= \left((\mu_A + \mu_B - 1_\Omega) \vee 0_\Omega, (\nu_A + \nu_B) \wedge 1_\Omega\right).$$

Also states can be extended to states on $\mathcal{M}$. Of course, you must define states on MV-algebras first.

Definition 2.7.4 *Let* $(G, \leq, 0, u, \oplus, \odot)$ *be a MV-algebra. A state on* $\mathcal{M}$ *is a mapping* $\bar{m}: \mathcal{M} \to [0,1]$ *satisfying the following conditions:*

(i) $\bar{m}(u) = 1, \bar{m}(0) = 0,$

(ii) $a \odot b = 0 \Rightarrow \bar{m}\big((a \oplus b)\big) = \bar{m}(a) + \bar{m}(b),$

(iii) $a_n \nearrow a \Rightarrow \bar{m}(a_n) \nearrow \bar{m}(a).$

Theorem 2.7.5 *Let* $\mathcal{M} \supset \mathcal{F}$ *be the MV-algebra constructed in Theorem 2.7.3. Then every state* $m: \mathcal{F} \to [0,1]$ *can be embedded to a state* $\tilde{m}: \mathcal{M} \to [0,1]$ *(*$\tilde{m} = m|\mathcal{F}$).*

Proof. It is easy to see that any element $(\mu_A, \nu_A) \in \mathcal{M}$ can be presented in the form

$$(\mu_A, \nu_A) \odot (0_\Omega, 1_\Omega - \nu_A) = (0_\Omega, 1_\Omega),$$

$$(\mu_A, 0_\Omega) = (\mu_A, \nu_A) \oplus (0_\Omega, 1_\Omega - \nu_A).$$

If $(\mu_A, \nu_A) \in \mathcal{F}$, then

$$m\big((\mu_A, 0_\Omega)\big) = m\big((\mu_A, \nu_A)\big) + m\big((0_\Omega, 1_\Omega - \nu_A)\big).$$

Generally, we can define $\bar{m}: \mathcal{M} \to [0,1]$ by the formula

$$\bar{m}\big((\mu_A, \nu_A)\big) = m\big((\mu_A, 0_\Omega)\big) - m\big((0_\Omega, 1_\Omega - \nu_A)\big),$$

so $\bar{m}$ is an extension of m. Of course, we must prove that $\bar{m}$ is a state. First we prove that $\bar{m}$ is additive.

Let $A = (\mu_A, \nu_A) \in \mathcal{M}$, $B = (\mu_B, \nu_B) \in \mathcal{M}$, $A \odot B = (0_\Omega, 1_\Omega)$, hence

$$\big((\mu_A + \mu_B - 1_\Omega) \vee 0_\Omega, (\nu_A + \nu_B) \wedge 1_\Omega\big) = (0_\Omega, 1_\Omega),$$

$$\mu_A + \mu_B \leq 1_\Omega, \nu_A + \nu_B \leq 1_\Omega.$$

Therefore

$$A \oplus B = \big((\nu_A + \mu_B) \wedge 1_\Omega, (\nu_A + \nu_B - 1_\Omega) \vee 0_\Omega\big) =$$

$$= (\nu_A + \mu_B, \nu_A + \nu_B - 1_\Omega).$$

By Theorem 2.3.6 we have

$$\bar{m}(A) = m(\mu_A, 0_\Omega) - m(0_\Omega, 1_\Omega - \nu_A) =$$

$$= \int \mu_A dP + \alpha(1 - \int (\mu_A + 0_\Omega)dQ) -$$

$$- \int 0_\Omega dP + \alpha(1 - \int (0_\Omega + 1_\Omega - \nu_A)dQ) =$$

$$= \int \mu_A dP + \alpha(1 - \int (\mu_A + \nu_A)dQ),$$

$$\bar{m}(B) = \int \mu_B dP + \alpha(1 - \int (\mu_B + \nu_B)dQ).$$

Hence also

$$\bar{m}(A \oplus B)$$

$$= \int (\mu_A + \mu_B)dP + \alpha(1 - \int (\mu_A + \mu_B + \nu_A + \nu_B - 1_\Omega)dQ) =$$

$$= \int (\mu_A + \mu_B)dP + \alpha(2 - \int (\mu_A + \mu_B + \nu_A + \nu_B)dQ) =$$

$$= \bar{m}(A) + \bar{m}(B).$$

Now we prove that $\bar{m}$ is monotone. Let $A \leq B$, *i.e.* $\mu_A \leq \mu_B$, $\nu_A \geq \nu_B$. Then

$$\bar{m}(B) - \bar{m}(A) = \int \mu_B dP + \alpha(1 - \int (\mu_B + \nu_B)dQ) -$$

$$- \int \mu_A dP + \alpha(1 - \int (\mu_A + \nu_A)dQ) =$$

$$= \int (\mu_B - \mu_A)dP - \alpha \int (\mu_B - \mu_A)dQ + \alpha \int (\nu_A - \nu_B)dQ \geq$$

$$\geq \int f dP - \alpha \int f dQ,$$

where $1_\Omega \geq f = \mu_B - \mu_A \geq 0$. If we consider in Theorem 2.3.6 the IF-set $(\chi_C, 0_\Omega)$, then we obtain

$$0 \leq \alpha = m(0_\Omega, 1_\Omega) \leq m(\chi_C, 0_\Omega) = P(C) + \alpha(1 - Q(C))$$

$$0 \leq P(C) - \alpha Q(C),$$

hence

$$0 \le \int f dP - \alpha \int f dQ,$$

for any non-negative function f.

Finally, let $A_n = \left(\mu_{A_n}, \nu_{A_n}\right) \in \mathcal{M}$, $A_n \nearrow A = (\mu_A, \nu_A)$, i.e. $\mu_{A_n} \nearrow \mu_A$, $\nu_{A_n} \searrow \nu_A$. Then

$$\bar{m}(A_n) = \int_\Omega \mu_{A_n} dP - \alpha \int_\Omega \mu_{A_n} dQ + \alpha - \alpha \int_\Omega \nu_{A_n} dQ \nearrow$$

$$\nearrow \int_\Omega \mu_A dP - \alpha \int_\Omega \mu_A dQ + \alpha - \alpha \int_\Omega \nu_A dQ = \bar{m}(A).$$

CHAPTER 3

Fuzzy Quantum Space

Abstract: In this chapter we study the existence of a sum of fuzzy observables in a fuzzy quantum space which generalizes the Kolmogorov probability space using the ideas of fuzzy set theory. We also study some properties of the sum of fuzzy observables. To study the above mentioned, we also include the basic notions from the probability theory on fuzzy quantum space in this chapter, *i.e.* the notion of fuzzy quantum space, a fuzzy observable, an indicator of a fuzzy set, a null fuzzy observable, a Boolean algebra on fuzzy quantum space, fuzzy state *etc.*

Keywords: Fuzzy quantum space, Fuzzy observable, Compatible fuzzy observable, Fuzzy state, Sum of fuzzy observables, Sum of non-compatible fuzzy observables, Properties of the sum of fuzzy observables, Convergences of fuzzy observables, Method of F-σ-ideals.

3.1 INTRODUCTION

In this chapter we firstly present the basic notions from the probability theory on fuzzy quantum spaces and we also introduce some properties of elements of a fuzzy quantum space, such as orthogonality and compatibility. Our studies are motivated by analogous notions that are intensively investigated in the theory of quantum logics Varadarajan [103, 104]. These notions have meaning not only in terms the probability theory, but they also have a physical motivation. For example, physical quantities can be represented by fuzzy observables. The introduction of the notion of compatibility has also a physical motivation, since non compatible observables correspond with those physical variables that do not admit measurability of their values.

Definition 3.1.1 *A fuzzy quantum space is a couple* $(\mathbb{X}, \mathcal{M})$*, where* $\mathbb{X}$ *is a nonempty set and* $\mathcal{M} \subset [0,1]^{\mathbb{X}}$ *satisfies the following conditions:*

(i) if $1_{\mathbb{X}}(x) = 1$ *for any* $x \in \mathbb{X}$*, then* $1_{\mathbb{X}} \in \mathcal{M}$*;*

(ii) if $f \in \mathcal{M}$*, then* $f' := 1_{\mathbb{X}} - f \in \mathcal{M}$

(iii) $\bigvee_{n=1}^{\infty} f_n = \sup_n f_n \in \mathcal{M}$*, for any* $\{f_n\}_{n=1}^{\infty} \subset \mathcal{M}$*;*

(iv) if $\frac{1}{2_{\mathbb{X}}}(x) = \frac{1}{2}$ *for any* $x \in \mathbb{X}$*, then* $\frac{1}{2_{\mathbb{X}}} \notin \mathcal{M}$*.*

Renáta Bartková, Beloslav Riečan and Anna Tirpáková

Elements of the set $\mathcal{M}$ are called fuzzy subsets of the universe $\mathbb{X}$. In particular, if f is the characteristic function, we call it a crisp set. The symbol $\bigvee_{n=1}^{\infty} f_n := \sup_n f_n$, respectively $\bigwedge_{n=1}^{\infty} f_n := \inf_n f_n$ indicates a fuzzy conjunction respectively a fuzzy disjunction of the sequence of fuzzy sets $\{f_n\}_{n=1}^{\infty} \subset \mathcal{M}$. The event $f' := 1_{\mathbb{X}} - f \in \mathcal{M}$ is the so-called a fuzzy negative.

In the fuzzy set theory the system $\mathcal{M}$ is called a soft σ-algebra [70]. The structure of fuzzy quantum space has been suggested by Riečan [79] as an alternative axiomatic model for quantum mechanics.

More general structure assuming that $\mathcal{M}$ is closed with respect to the union of any sequence of mutually orthogonal fuzzy sets has been proposed by Pykacz [73] and studied by Dvurečenskij and Chovanec [17]. Some fuzzy sets ideas have been studied also by Guz [30], but his approach is different from ours.

Let $(\mathbb{X}, \mathcal{M})$ be a fuzzy quantum space. The set $\mathcal{M}$ may be regarded as a partially ordered set in which we define $f \leq g$ if and only if $f(x) \leq g(x)$ for any element $x \in \mathbb{X}$. Using the complementation $': f \to f' = 1_{\mathbb{X}} - f$ for any fuzzy set $f \in \mathcal{M}$, we see that the complementation $'$ satisfies two conditions:

(i) $(f')' = f$ for any $f \in \mathcal{M}$;

(ii) if $f \leq g$, then $g' \leq f'$.

It is evident that $f \vee f' = 1_{\mathbb{X}}$ if and only if f is a crisp set. Hence $\mathcal{M}$ is a distributive σ-lattice with complementation $'$ for which de Morgan laws hold

$$\left(\bigvee_{n=1}^{\infty} f_n \right)' = \bigwedge_{n=1}^{\infty} f_n',$$

$$\left(\bigwedge_{n=1}^{\infty} f_n \right)' = \bigvee_{n=1}^{\infty} f_n'$$

for any sequence $\{f_n\}_{n=1}^{\infty} \subset \mathcal{M}$.

Remark 3.1.2 *Let* $(\mathbb{X}, \mathcal{M})$ *be a fuzzy quantum space. In the system $\mathcal{M}$ there are two special classes of sets:*

1. $W_0(\mathcal{M})$ *denotes the set of all W-empty sets, i.e. those* $f \in \mathcal{M}$ *for which* $f(x) \leq \frac{1}{2}$ *for every* $x \in \mathbb{X}$;

2. $W_1(\mathcal{M})$ *denotes the set of all W-universums, i.e. those* $g \in \mathcal{M}$ *for which* $g(x) \geq \frac{1}{2}$ *for every* $x \in \mathbb{X}$;

It is clear that for every fuzzy set $f \in \mathcal{M}$, *the set* $f \wedge f' \in W_0(\mathcal{M})$ *and* $f \vee f' \in W_1(\mathcal{M})$.

The analogue of a random variable is a fuzzy quantum observable, which we define as follows.

Definition 3.1.3 *A fuzzy observable on a fuzzy quantum space* $(\mathbb{X}, \mathcal{M})$ *is a mapping* $x: \mathcal{B}(\mathbb{R}^1) \to \mathcal{M}$ *satisfying the following properties:*

(i) $x(E^c) = 1_{\mathbb{X}} - x(E)$ *for every* $E \in \mathcal{B}(\mathbb{R}^1)$,

(ii) if $\{E_n\}_{n=1}^{\infty} \subset \mathcal{B}(\mathbb{R}^1)$, *then* $x(\bigcup_{n=1}^{\infty} E_n) = \bigvee_{n=1}^{\infty} x(E_n)$,

where $\mathcal{B}(\mathbb{R}^1)$ *denotes the Borel* σ-*algebra of the real line* $\mathbb{R}^1$ *and* E^c *denotes the complement of a set* E *in* $\mathbb{R}^1$.

Definition 3.1.4 *Let* $f \in \mathcal{M}$. *The mapping* $x_f: \mathcal{B}(\mathbb{R}^1) \to \mathcal{M}$ *defined by*

$$x_f(E) = \begin{cases} f \wedge f', & \text{if } 0,1 \notin E, \\ f', & \text{if } 0 \in E, 1 \notin E, \\ f, & \text{if } 0 \notin E, 1 \in E, \\ f \vee f', & \text{if } 0,1 \in E, \end{cases}$$

for every $E \in \mathcal{B}(\mathbb{R}^1)$ *is a fuzzy observable of a fuzzy quantum space* $(\mathbb{X}, \mathcal{M})$ *called the indicator of a fuzzy quantum set* $f \in \mathcal{M}$.

It is evident that $x_f(E)$ plays the role of an indicator of the fuzzy set $f \in \mathcal{M}$.

Remark 3.1.5 *If* $\tau: \mathbb{R}^1 \to \mathbb{R}^1$ *is a Borel measurable function and* x *is a fuzzy observable, then*

$$\tau \circ x: E \to x(\tau^{-1}(E)), E \in \mathcal{B}(\mathbb{R}^1),$$

is a fuzzy observable, too. In this way, we can define the functional calculus of observables. For example, if $\tau(t) = t^2$, $t \in \mathbb{R}^1$, we write $\tau \circ x = x^2$ and the like. In particular, if $a \in \mathbb{R}^1$, then $ax: E \rightarrow x(\{t \in \mathbb{R}^1: at \in E\})$ for any $E \in \mathcal{B}(\mathbb{R}^1)$.

Remark 3.1.6 *The spectrum of a fuzzy observable x we mean the set $\sigma(x) =$ $\cap \ \{C \subset \mathbb{R}^1: C \ is closed and \ x(C) = x(\mathbb{R}^1)\}$. A fuzzy observable x is bounded if $\sigma(x)$ is a bounded set, in this case, we define the norm $\|x\|$ of the observable x, via*

$$\|x\| = \sup\{|t|: t \in \sigma(x)\}.$$

The question observable of the null fuzzy set $0_{\mathbb{X}}$ we denote by o and we define as follows.

Definition 3.1.7 *The null fuzzy observable is a mapping $o: \mathcal{B}(\mathbb{R}^1) \rightarrow \mathcal{M}$ defined by*

$$o(E) = \begin{cases} 0_{\mathbb{X}}, & if \ 0 \notin E, \\ 1_{\mathbb{X}}, & if \ 0 \in E, \end{cases}$$

where $E \in \mathcal{B}(\mathbb{R}^1)$.

We will define a Boolean σ-algebra $\mathcal{A}$ for a fuzzy quantum space $(\mathbb{X}, \mathcal{M})$ as follows.

Definition 3.1.8 *A nonempty subset $\mathcal{A} \subset \mathcal{M}$ is called a Boolean algebra (σ-algebra) of a fuzzy quantum space $(\mathbb{X}, \mathcal{M})$ if*

(i) there are minimal and maximal elements $0_{\mathcal{A}}$ and $1_{\mathcal{A}}$ from $\mathcal{A}$ such that for any $f \in \mathcal{A}$, $0_{\mathcal{A}} \leq f \leq 1_{\mathcal{A}}$ and $f \vee f' = 1_{\mathcal{A}}$ (we recall that $0_{\mathcal{A}}$ and $1_{\mathcal{A}}$ are not crisp sets, in general);

(ii) $\mathcal{A}$ is with respect to $\wedge, \vee, 0_{\mathcal{A}}, 1_{\mathcal{A}}$ a Boolean algebra (σ-algebra).

It is clear that $0_{\mathcal{A}} \neq 1_{\mathcal{A}}$. For example, if f is a fuzzy set from $\mathcal{M}$, then $\mathcal{A} = \{f \wedge f', f, f', f \vee f'\}$ is a Boolean algebra with the minimal and maximal elements $0_{\mathcal{A}_f} = f \wedge f'$ and $1_{\mathcal{A}_f} = f \vee f'$, respectively.

In particular, if x is a fuzzy observable of a fuzzy quantum space $(\mathbb{X}, \mathcal{M})$, then the range $\mathcal{R}(x) = \{x(E): E \in \mathcal{B}(\mathbb{R})\}$ is a Boolean σ-algebra of $(\mathbb{X}, \mathcal{M})$ with the minimal and maximal elements $0_{\mathcal{R}(x)} = x(\emptyset)$ and $1_{\mathcal{R}(x)} = x(\mathbb{R}^1)$.

In accordance with the theory of quantum logics, we say that two fuzzy sets $f, g \in \mathcal{M}$ are:

(i) orthogonal, if $f \leq 1 - g$, and we write $f \perp g$;

(ii) compatible, if $f = (f \wedge g) \vee (f \wedge g')$, $g = (g \wedge f) \vee (g \wedge f')$ and write $f \leftrightarrow g$;

(iii) strongly compatible, if $f \leftrightarrow g \leftrightarrow f' \leftrightarrow g'$, and we write $f \overset{s}{\leftrightarrow} g$.

Two fuzzy observables x and y are compatible, if $x(E) \leftrightarrow y(F)$ for any $E, F \in \mathcal{B}(\mathbb{R}^1)$.

The following example demonstrates when two fuzzy sets of a fuzzy quantum space $(\mathbb{X}, \mathcal{M})$ are compatible, but do not strongly compatible.

Example 3.1.9 *Let f, g be two constant functions from $\mathcal{M}$, where $f = 0.2$, $g = 0.3$, $f' = 0.8$ and $g' = 0.7$, then $f \leftrightarrow g$, but $f \overset{s}{\nleftrightarrow} g$.*

The basic concept of the theory of a quantum logic and a probability theory is a state or a probability measure. In our case, we will define the fuzzy state as follows.

Definition 3.1.10 *A fuzzy state on a fuzzy quantum space $(\mathbb{X}, \mathcal{M})$ is a mapping $m: \mathcal{M} \to [0,1]$ such that*

(i) $m(f \vee (1_{\mathbb{X}} - f)) = 1$ for every $f \in \mathcal{M}$;

(ii) if $\{f_k\}_{k=1}^{\infty}$ is a sequence of pairwise orthogonal fuzzy subsets from $\mathcal{M}$, i.e. $f_i \perp f_j$, $(f_i \leq 1_{\mathbb{X}} - f_j)$, whenever $i \neq j$, then

$$m \left(\overset{\infty}{\underset{k=1}{\vee}} f_i \right) = \Sigma_{k=1}^{\infty} m(f_i).$$

According to Piasecki [135], a fuzzy state is called the P-measure. The triplet $(\mathbb{X}, \mathcal{M}, m)$ where m is a P-measure, is called a fuzzy probability space. This structure was studied, *e.g.* in Markechová ([44, 45]).

From the following example it follows that the notions generalize the component notions of classical measure theory [33].

Example 3.1.11 *[50] Let $(\mathbb{X}, S)$ be a measurable space in the classical sense, i.e. S is a σ-algebra of crisp subsets of $\mathbb{X}$. Put $\mathcal{M} = \{\chi_A; A \in S\}$ where χ_A is the indicator of the set $A \in S$. Then the couple $(\mathbb{X}, \mathcal{M})$ is a fuzzy quantum space. If $v: S \to \mathbb{R}$ is a σ-additive function, then it is easy verify that the mapping $m: \mathcal{M} \to \mathbb{R}$ defined by $m(\chi_A) = v(A)$ is a fuzzy state of $(\mathbb{X}, \mathcal{M})$. If $(\mathbb{X}, S, v)$ is a probability space in the sense of classical probability theory, then the triplet $(\mathbb{X}, \mathcal{M}, m)$ is a fuzzy probability space.*

For illustration, we give a following example of a nontrivial fuzzy quantum space [44].

Example 3.1.12 *Let $(\mathbb{X}, \mathcal{M})$ where $\mathbb{X} = [0,1]$, $f: \mathbb{X} \to \mathbb{X}$, $f(x) = x$, $\mathcal{M} = \{f, f', f \vee f', f \wedge f', 0_{\mathbb{X}}, 1_{\mathbb{X}}\}$ for every $x \in \mathbb{X}$. It is evident that $f \vee f' \neq 1_{\mathbb{X}}$. We define the mapping $m: \mathcal{M} \to [0,1]$ by the equalities $m(1_{\mathbb{X}}) = m(f \vee f') = 1$, $m(0_{\mathbb{X}}) = m(f \wedge f') = 0$ and $m(f) = m(f') = \frac{1}{2}$, then the triplet $(\mathbb{X}, \mathcal{M}, m)$ is a fuzzy probability space.*

3.2 SUM OF FUZZY QUANTUM OBSERVABLES

In the following section we study the existence of a sum of fuzzy observables of a fuzzy quantum space. The sum of fuzzy observables for a compatible case was for the first time defined by Harman and Riečan [34] and this form of a sum was used, e.g. in Palumbíny's work [66].

In this section, we show that the sum of fuzzy observables exists also for non-compatible fuzzy observables, and it is determined uniquely. We also introduce several selected properties of the sum of fuzzy observables here.

At first we introduce the theorem, which is further be used for the properties of fuzzy observables of a fuzzy quantum space. This theorem has been proved by Dvurečenskij and Riečan [23].

Theorem 3.2.1 *Let $\{f_t: t \in T\}$ be a system of fuzzy sets from $\mathcal{M}$. The following assertions are equivalent:*

(i) $\{f_t: t \in T\}$ is a system of mutually strongly compatible elements;

(ii) $f_s \vee f'_s = f_t \vee f_{t'}$ for any $s, t \in T$;

(iii) there is a Boolean σ-algebra of $\mathcal{M}$ containing all $\{f_t: t \in T\}$.

Now we characterize fuzzy observables of a fuzzy quantum space $(\mathbb{X}, \mathcal{M})$.

Theorem 3.2.2 *Let x be a fuzzy observable of a fuzzy quantum space $(\mathbb{X}, \mathcal{M})$ and let $B_x(t) = x((-\infty, t))$, $t \in \mathbb{R}^1$. Then system $\{B_x(t): t \in \mathbb{R}^1\}$ satisfies the following conditions:*

(i) $B_x(s) \leq B_x(t)$ if $s \leq t$;

(ii) $\underset{t}{\vee} B_x(t) = f$;

(iii) $\underset{t}{\wedge} B_x(t) = f'$;

(iv) $\underset{t<s}{\vee} B_x(t) = B_x(s)$;

(v) $B_x(t) \vee B_x'(t) = f$, where $f = x(\mathbb{R}^1)$ and $f' = x(\emptyset)$.

Conversely, if a system $\{B(t): t \in \mathbb{R}^1\}$ of fuzzy sets of a fuzzy quantum space $(\mathbb{X}, \mathcal{M})$ meets conditions (i)-(v) for an $f \in \mathcal{M}$, then there is a unique fuzzy observable x such that $B_x(t) = B(t)$ for any t, and $x(\mathbb{R}^1) = f$.

Proof.

(i) If $s \leq t$, then for $(-\infty, t) \cup [s, t)$ we have

$$x((-\infty, t)) = x((-\infty, t)) \vee x([s, t)),$$

it follows that

$$B_x(t) = x((-\infty, t)) \geq x((-\infty, s)) = B_x(s).$$

(ii) If $f = x(\mathbb{R}^1)$, then $x((\infty, t)) \leq f$ for every $t \in \mathbb{R}^1$. For every integer n we have $x((\infty, n)) \leq f$ and

$$x(\mathbb{R}^1) = x(\cup_{n=1}^{\infty} (-\infty, n)) = \overset{\infty}{\underset{n=1}{\vee}} x((-\infty, n)).$$

(iii) Since $x((-\infty, n)) \geq x(\emptyset)$ for every integer n, we have

$$x(\emptyset) = x(\cap_{n=1}^{\infty} (-\infty, n)) = \overset{\infty}{\underset{n=1}{\wedge}} ((-\infty, n)).$$

(iv) An interval $(-\infty, s)$ can be written as follows:

$$(-\infty, s) = \bigcup_{n=1}^{\infty} \left(-\infty, s - \frac{1}{n}\right),$$

then for the fuzzy observable x we have

$$x(-\infty, s) = x\left(\bigcup_{n=1}^{\infty} \left(-\infty, s - \frac{1}{n}\right)\right) = \bigvee_{n=1}^{\infty} x\left(\left(-\infty, s - \frac{1}{n}\right)\right) =$$

$$= \bigvee_{t<s} x((-\infty, t)).$$

(v) It may be proved as follows:

$$B_x(t) \vee B_x'(t) = x((-\infty, t) \cup (-\infty, t)^c) = x(\mathbb{R}^1).$$

Conversely. For the fuzzy quantum space $(\mathbb{X}, \mathcal{M})$ let now a system $\{B(t): t \in \mathbb{R}^1\}$ satisfying conditions *(i)–(v)* be given. Due to *(v)*, the system $\{B(t): t \in \mathbb{R}^1\}$ consists of mutually strongly compatible elements of $\mathcal{M}$, so that, according to Theorem 3.2.1, there is a minimal Boolean σ-algebra $\mathcal{A}$ of $\mathcal{M}$ containing all $B(t)$. By the Loomis-Sikorski theorem [97], there is a measurable space $(\Omega, \mathcal{S})$ and a σ-homomorphism h from $\mathcal{S}$ onto $\mathcal{A}$.

Let $r_1, r_2, r_3, \dots$ be any distinct enumeration of rational numbers. We claim to construct by induction the sets $A_1, A_2, A_3 \dots$ from $\mathcal{S}$ such that

$$h(A_i) = B(r_i); \tag{3.1}$$

$$A_i \subset A_j \text{ if } r_i < r_j; \tag{3.2}$$

$$\bigcap_{i=1}^{\infty} A_i = \emptyset. \tag{3.3}$$

We note that if $A \subset B$, $A, B \in \mathcal{S}$ and if there is $c \in \mathcal{A}$ such that $h(A) \leq c \leq h(B)$ then there is $C \in \mathcal{S}$ such that $A \subset C \subset B$, $h(C) = c$. Indeed, since h maps $\mathcal{S}$ onto $\mathcal{A}$, there is a $C_1 \in \mathcal{S}$ such that $h(C_1) = c$. If we define $C = (C_1 \cap B) \cup A$ then C has the given property.

Let A_1 be any set in $\mathcal{S}$ such that $h(A_1) = B(r_1)$. Suppose $A_1, A_2, \dots, A_n \in \mathcal{S}$ have been constructed so that (3.1) and (3.2) hold.

We shall construct A_{n+1} as follows. Let $(i_1, i_2, \ldots, i_n)$ be the permutation of $(1, 2, \ldots, n)$ such that $r_{i_1} < r_{i_2} < \ldots < r_{i_n}$. Then only one of the following conditions holds:

$$r_{n+1} < r_{i_1}; \tag{3.4}$$

$$r_{n+1} > r_{i_n}; \tag{3.5}$$

there is a unique $k = 1, 2, \ldots, n - 1$ such that

$$r_{i_k} < r_{n+1} < r_{i_{k+1}}, \tag{3.6}$$

and by the above observation we can select A_{n+1} such that

$$h(A_{n+1}) = B(r_{n+1})$$

and

$$A_{n+1} \subseteq A_i; \tag{3.7}$$

$$A_{n+1} \supseteq A_{i_n}; \tag{3.8}$$

$$A_{i_k} \subseteq A_{n+1} \subseteq A_{i_{k+1}}, \tag{3.9}$$

according to the properties $(3.4) - (3.6)$. Then the system $\{A_1, A_2, \ldots, A_n\}$ fulfills properties (3.1) and (3.2). Thus, by induction, it follows that there is a sequence $\{A_j\}_{j=1}^{\infty}$ of sets in $\mathcal{S}$ with properties (3.1) and (3.2). As

$$h\left(\cap_{j=1}^{\infty} A_j\right) = \overset{\infty}{\underset{j=1}{\wedge}} h(A_j) = \overset{\infty}{\underset{j=1}{\wedge}} B(r_j) = 0_{\mathcal{A}},$$

we may replacing A_j by $A_j - \cap_i A_i$ if necessary, assume that $\cap_i A_i = \emptyset$. We define an $\mathcal{S}$-measurable function u as follows:

$$u(\omega) = \begin{cases} 0, \text{if } \omega \notin \cup_{j=1}^{\infty} A_j \\ \inf\{r_j \colon \omega \in A_j\}, \text{if } \omega \in \cup_{j=1}^{\infty} A_j. \end{cases}$$

The function u is everywhere well-defined and it is finite. Moreover,

$$u^{-1}\big((-\infty, r_k)\big) = \begin{cases} \bigcup_{r_j < r_k} A_j, \text{if } r_k \leq 0, \\ \bigcup_{r_j < r_k} A_j \cup (\Omega - \bigcup_i A_i), \text{if } r_k > 0, \end{cases}$$

hence u is S-measurable and $h(u^{-1}((-\infty, r_k))) = B(r_k)$. If we define an observable by $x(E) = h(u^{-1}(E))$, $E \in \mathcal{B}(\mathbb{R}^1)$, then $x((-\infty, t)) = B(t)$ for every $t \in \mathbb{R}^1$. The equality $x_1((-\infty, t)) = x_2((-\infty, t))$ for every $t \in \mathbb{R}^1$ implies $x_1 = x_2$, hence, the uniqueness of x is shown and the proof is complete.

If u and v are two S-measurable functions, then the measurability of the sum $u + v$ can be proved using the following simple relation

$$\{k \in \mathbb{X} : (u + v)(k) < t\} = \bigcup_{r \in Q} (\{k \in \mathbb{X} : u(k) < r\} \cap \{k \in \mathbb{X} : v(k) < t - r\}),$$

where Q is the set of all rational numbers. With this fact, we can define the sum of fuzzy observables similarly as Dvurečenskij defined the sum of the non-compatible observable on quantum logics in [15].

In accordance with this simple relation given above, we define sum of two observables as follows.

Definition 3.2.3 *Let x and y be two fuzzy observables of a fuzzy quantum space* $(\mathbb{X}, \mathcal{M})$. *If the system* $\{B_{x+y}(t) : t \in \mathbb{R}^1\}$,

$$B_{x+y}(t) = \bigvee_{r \in Q} \big(B_x(r) \wedge B_y(t - r)\big), t \in \mathbb{R}^1,$$

where Q is the set of all rational, determines a fuzzy observable z of $(\mathbb{X}, \mathcal{M})$, *then we call it the sum of x and y, and we write $y = x + y$.*

It is clear that if the sum exists, then it is unique. In the following we will introduce lemma, which we use in the proof of Theorem 3.2.5.

Lemma 3.2.4 *Let S be a dense countable set in $\mathbb{R}^1$. For fuzzy observables x and y let us denote*

$$B_{x+y}^S(t) = \bigvee_{s \in S} \big(B_x(s) \wedge B_y(t - s)\big),$$

then

$$B_{x+y}^S(t) = B_{x+y}(t)$$

for every $t \in \mathbb{R}^1$.

Proof. We can show that if t_n converges to t (denote $t_n Z t$), $t_n \in S, t \in \mathbb{R}^1$, then $B_{x+y}^S(t) = \bigvee_n B_{x+y}^S(t_n)$. Indeed,

$$\bigvee_n B_{x+y}^S(t_n) = \bigvee_n \bigvee_{s \in S} \left(B_x(s) \wedge B_y(t_n - s) \right) =$$

$$= \bigvee_{s \in S} \left(B_x(s) \wedge \bigvee_n B_y(t_n - s) \right) =$$

$$= \bigvee_{s \in S} \left(B_x(s) \wedge B_y(t - s) \right).$$

Let now n be any integer, then for each $s \in S$ there is $r = r(s) \in Q$ such that we have $s < r < s + \frac{1}{n}$. Therefore,

$$B_x(s) \wedge B_y(t - n^{-1} - s) \leq B_x(r) \wedge B_y(t - r)$$

and

$$B_{x+y}^S(t - n^{-1}) \leq B_{x+y}(t),$$

$$B_{x+y}^S(t) \leq \bigvee_{s \in S} B_{x+y}^S(t - n^{-1}) \leq B_{x+y}(t).$$

Similarly we can show that $B_{x+y}(t) \leq B_{x+y}^S(t)$.

Theorem 3.2.5 *For every two fuzzy observables x and y of a fuzzy quantum space $(\mathbb{X}, \mathcal{M})$ their sum exists.*

Proof. We show that the system $\{B_{x+y}(t): t \in \mathbb{R}^1\}$ fulfills the conditions of Theorem 3.2.2. The proof of (i), (ii) and (iv) (Theorem 3.2.2) is simple, due to the σ-continuity of $\mathcal{M}$, that is, if $f_i \in \mathcal{M}, i = 1,2,\dots, f_1 \leq f_2 \leq \dots$, then for any $g \in \mathcal{M}$,

$$g \wedge \left(\bigvee_i f_i \right) = \bigvee_i (g \wedge f_i).$$

(i) The interval $(-\infty, t)$ can be written as follows: $(-\infty, t) = (-\infty, s) \cup [s, t)$ for $s \leq t$, where $s, t \in \mathbb{R}^1$. Then

$$B_{x+y}(t) = \bigvee_{r \in Q} \left(B_x(r) \wedge B_y(t - r) \right) \geq \bigvee_{r \in Q} \left(B_x(r) \wedge B_y(s - r) \right) =$$

$$= B_{x+y}(s).$$

(ii) Denote $f = x(\mathbb{R}^1) \wedge y(\mathbb{R}^1)$. Then

$$\bigvee_t B_{x+y}(t) = \bigvee_t \bigvee_{r \in Q} \left(B_x(r) \wedge B_y(t - r) \right) =$$

$$= \bigvee_{r \in Q} \bigvee_t \left(B_x(r) \wedge B_y(t - r) \right) =$$

$$= \bigvee_{r \in Q} B_x(r) \wedge \bigvee_t B_y(t - r) = \bigvee_{r \in Q} B_x(r) \wedge y(\mathbb{R}^1) =$$

$$= x(\mathbb{R}^1) \wedge y(\mathbb{R}^1) = f.$$

(iv) For each $s \in \mathbb{R}^1$ the interval

$$(-\infty, s) = \bigcup_{n=1}^{\infty} \left(-\infty, s - \frac{1}{n} \right).$$

Then

$$B_{x+y}(s) = \bigvee_{r \in Q} \left(B_x(r) \wedge B_y(s - r) \right) = \bigvee_{r \in Q} B_x(r) \wedge \bigvee_{n=1}^{\infty} B_y \left(s - \frac{1}{n} - r \right) =$$

$$= \bigvee_{n=1}^{\infty} \left(\bigvee_{r \in Q} \left(B_x(r) \wedge B_y \left(s - \frac{1}{n} - r \right) \right) \right) = \bigvee_{n=1}^{\infty} B_{x+y} \left(s - \frac{1}{n} \right).$$

(v) Calculate

$$B_{x+y}(t) \vee B'_{x+y}(t) = \bigvee_{r \in Q} \left(B_x(r) \wedge B_y(t - r) \right) \vee \bigwedge_{s \in Q} \left(B'_x(s) \vee B'_y(t - s) \right) =$$

$$= \bigwedge_{s} \bigvee_{r} \left(B_x(r) \wedge B_y(t - r) \right) \vee \left(B'_x(s) \vee B'_y(t - s) \right) =$$

$$= \bigwedge_{s} \bigvee_{r} \left(B_x(r) \vee \left(B'_x(s) \vee B'_y(t - s) \right) \right) \wedge$$

$$\wedge \left(B_y(t - r) \vee \left(B'_x(s) \vee B'_y(t - s) \right) \right).$$

Since

$$B_x(r) \vee B'_x(r) = x(\mathbb{R}^1) \text{ and } B_x(r) \vee B'_x(s) = x(\mathbb{R}^1) \text{ for } s \leq r, \text{ we have}$$

$$B_{x+y}(t) \vee B'_{x+y}(t) =$$

$$=\bigwedge_s \left(\bigvee_{r\geq s} \left(\left(B_x(r) \vee B'_x(s) \vee B'_y(t-r)\right) \wedge \left(B_y(t-r) \vee B'_x(s) \vee B'_y(t-s)\right)\right) \vee \right.$$

$$\left. \vee \bigvee_{r<s} \left(\left(B_x(r) \vee B'_x(s) \vee B'_y(t-s)\right) \wedge \left(B_y(t-r) \vee B'_x(s) \vee B'_y(t-s)\right)\right)\right) =$$

$$=\bigwedge_s \left(\left(\left((x(\mathbb{R}^1) \vee B'_y(t-s)) \wedge \left(\bigvee_{r\geq s} B_y(t-r) \vee B'_x(s) \vee B'_y(t-s)\right)\right)\right) \vee \right.$$

$$\left. \vee \left(\bigvee_{r<s} \left(B_x(r) \vee B'_x(s) \vee B'_y(t-s)\right) \wedge \left(y(\mathbb{R}^1) \vee B'_x(s)\right)\right)\right),$$

Since $r \searrow s$ implies $t-r \nearrow t-s$ we have $\bigvee_{r\geq s} B_y(t-r) = B_y(t-s)$. Moreover, de Morgan rule gives that $\bigwedge_s B'(s) = (\bigvee B(s))'$. Then

$$B_{x+y}(t) \vee B'_{x+y}(t) =$$

$$=\bigwedge_s \left(\left(\left(x(\mathbb{R}^1) \vee B'_y(t-s)\right) \wedge (\bigvee_{r\geq s} B_y(t-r) \vee B'_x(s) \vee B'_y(t-s))\right)\right) \vee$$

$$\vee \left(\bigvee_{r<s} \left(\left(B_x(r) \vee B'_x(s) \vee B'_y(t-s)\right) \wedge (y(\mathbb{R}^1) \vee B'_x(s))\right)\right) =$$

$$=\bigwedge_s \left(\left((x(\mathbb{R}^1) \vee B'_y(t-s)) \wedge (y(\mathbb{R}^1) \vee B'_x(s))\right) \vee \right.$$

$$\vee \left(((x(\mathbb{R}^1) \vee B'_y(t-s)) \wedge (y(\mathbb{R}^1) \vee B'_x(s)))\right)) =$$

$$= \left(\left(x(\mathbb{R}^1) \vee (\bigvee_s B_y(t-s))'\right) \wedge (y(\mathbb{R}^1) \vee (\bigvee_s B_x(s))')\right) \vee$$

$$\vee \left(x(\mathbb{R}^1) \vee (\bigvee_s B_y(t-s))'\right) \wedge \left(y(\mathbb{R}^1) \vee (\bigvee_s B_x(s))'\right) =$$

$$= \left((x(\mathbb{R}^1) \vee y(\emptyset)) \wedge (y(\mathbb{R}^1) \vee x(\emptyset))\right) \vee$$

$$\vee \left((x(\mathbb{R}^1) \vee y(\emptyset)) \wedge (y(\mathbb{R}^1) \vee x(\emptyset))\right)$$

In other words, we have proved

$$B_{x+y}(t) \vee B'_{x+y}(t) = (x(\mathbb{R}^1) \vee y(\emptyset)) \wedge (y(\mathbb{R}^1) \vee x(\emptyset)) = x(\mathbb{R}^1) \wedge y(\mathbb{R}^1) = f,$$

which means the strong compatibility of $\{B_{x+y}(t): t \in \mathbb{R}^1\}$, too. To prove the property (*iii*) of Theorem 3.2.2

$$\bigwedge_t B_{x+y}(t) = f' = x(\emptyset) \vee y(\emptyset), \tag{3.10}$$

we take into account that, by virtue of the property (*v*) of Theorem 3.2.2, $\{B_{x+y}(t): t \in \mathbb{R}^1\}$ is a system of mutually strongly compatible elements of a fuzzy quantum space $(\mathbb{X}, \mathcal{M})$.

By Lemma 3.2.4, it suffices to prove (3.10) for $t \in T$, where T is a countable dense subset of $\mathbb{R}^1$. By Theorem 2.2, there exists a Boolean σ-algebra $\mathcal{A} \subset \mathcal{M}$ containing all $B_{x+y}(t)$ for any $t \in \mathbb{R}^1$. Every Boolean σ-algebra $\mathcal{A}$ of $\mathcal{M}$ is σ-distributive, that is, if T and S are countable sets, then

$$\bigwedge_{t \in T} \bigvee_{s \in S} f_{ts} = \bigvee_{\varphi \in S^T} \bigwedge_{t \in T} f_{t\varphi(t)}$$

for any two indexed sequences $\{f_{ts}: t \in T, s \in S\} \subset \mathcal{M}$.

In particular, by Sikorski [97] a Boolean σ-algebra $\mathcal{A}$ is σ-distributive iff for any $f \in \mathcal{A}$, $f \neq 0_{\mathcal{A}}$, and any sequence $\{f_n\} \subset \mathcal{A}$ there exists $\{e(n)\}_{n=1}^{\infty} \in \{0,1\}$ such that

$$f \wedge \bigwedge_{n=1}^{\infty} f_n^{e(n)} \neq 0_{\mathcal{A}},$$

where $f_n^0 = f'_n$, $f_n^1 = f_n$, which is easily verifiable in our case.

Then

$$\bigwedge_{t \in T} \bigvee_{r \in Q} \left(B_x(r) \wedge B_y(t - r)\right) = \bigvee_{\varphi \in Q^T} \bigwedge_{t \in T} \left(B_x(\varphi(t)) \wedge B_y(t - \varphi(t))\right). \tag{3.11}$$

It is clear that

$$\bigwedge_t B_{x+y}(t) \geq x(\emptyset) \vee y(\emptyset) = x(\mathbb{R}^1) \wedge (y(\emptyset) \vee x(\emptyset)) \wedge y(\mathbb{R}^1). \tag{3.12}$$

Let $\varphi \in Q^T$, then

$$\bigwedge_{t \in T} \Big(B_x(\varphi(t)) \wedge B_y(t - \varphi(t)) \Big) = \bigwedge_{t \in T} \Big(B_x(\varphi(t)) \wedge \bigwedge_{t \in T} B_y(t - \varphi(t)) \Big).$$

There are two possible cases:

(a) $\inf_{t \in T} \varphi(t) = k > -\infty$, then

$$\bigwedge_{t \in T} B_x(\varphi(t)) \wedge \bigwedge_{t \in T} B_y(t - \varphi(t)) = \bigwedge_{t \in T} \Big(B_x(\varphi(t)) \wedge \bigwedge_{t \in T} B_y(t - \varphi(t)) \Big) =$$

$$= B_x(k) \wedge \bigwedge_{t \in T} B_y(t - \varphi(t)) =$$

$$= B_x(k) \wedge y(\emptyset) \le x(\mathbb{R}^1) \wedge y(\emptyset) \le$$

$$\le x(\emptyset) \vee y(\emptyset).$$

(b) $\inf_{t \in T} \varphi(t) = -\infty$, then

$$\bigwedge_{t \in T} B_x \Big(\varphi(t) \wedge \bigwedge_{t \in T} B_y(t - \varphi(t)) \Big) = x(\emptyset) \wedge y(\mathbb{R}^1) \le x(\emptyset) \vee y(\emptyset).$$

For every $\varphi \in Q^T$, we have

$$\bigwedge_{t \in T} B_x \Big(\varphi(t) \wedge \bigwedge_{t \in T} B_y(t - \varphi(t)) \Big) \le x(\emptyset) \vee y(\emptyset),$$

and taking into account (4.11) and (4.12), the the following inequalities hold:

$$x(\emptyset) \vee y(\emptyset) \le \bigwedge_{t \in T} B_{x+y}(t) \le x(\emptyset) \vee y(\emptyset).$$

Evidently, for any $t \in \mathbb{R}^1$, $B_{x+y}(t) \ge x(\emptyset) \vee y(\emptyset)$. Therefore

$$x(\emptyset) \vee y(\emptyset) = \bigwedge_{t \in T} B_{x+y}(t) \ge \bigwedge_{t \in \mathbb{R}^1} B_{x+y}(t) \ge x(\emptyset) \vee y(\emptyset).$$

Properties of the Sum of Fuzzy Observables

In this part, we establish some of the basic properties of the sum of observables.

We recall that if $x \leftrightarrow y$, then, according by Dvurečenskij and Riečan [23], there is a fuzzy observable z and two Borel measurable functions u and v such that $x = u \circ z$, $y = v \circ z$.

Some selected properties of the sum of the fuzzy observables are summarized in the following theorem.

Theorem 3.2.6

(i) $x + y = y + x$ for any two fuzzy observables x and y.

(ii) $(x + y) + z = x + (y + z)$ for any three fuzzy observables x, y and z.

(iii) If $x \leftrightarrow y$, then $x + y = (u + v) \circ z$ provided $x = u \circ z$, $y = v \circ z$.

(iv) Let $\alpha \in \mathbb{R}^1$ and put

$$I_\alpha(E) = \begin{cases} 1_{\mathbb{X}}, \text{if } \alpha \in E, \\ 0_{\mathbb{X}}, \text{if } \alpha \notin E, (E \in \mathcal{B}(\mathbb{R}^1)) \end{cases}$$

then $x + I_\alpha = f_\alpha \circ x$, where $f_\alpha(t) = t + \alpha$.

(v) $\beta(x + y) = \beta x + \beta y$ for any $\beta \in \mathbb{R}^1$ and every fuzzy observables x and y.

Proof.

(i) Let $t \in \mathbb{R}^1$ and denote $S_t = \{t - r : r \in Q\}$. Then S_t is dense in $\mathbb{R}^1$ and using Lemma 3.2.4, we have

$$B_{x+y}(t) = \bigvee_{r \in Q} \left(B_x(r) \wedge B_y(t - r)\right) =$$

$$= \bigvee_{s \in S_t} \left(B_y(s) \wedge B_x(t - s)\right) =$$

$$= B_{x+y}^S(t) = B_{y+x}(t).$$

(ii)
$$B_{(x+y)+z}(t) = \bigvee_{r \in Q} \left(B_{x+y}(r) \wedge B_z(t - r)\right) =$$

$$= \underset{r \in Q}{\vee} \left(\underset{s \in Q}{\vee} \left(B_x(s) \wedge B_y(r - s) \right) \wedge B_z(t - r) \right) =$$

$$= \underset{s \in Q}{\vee} B_x(s) \wedge \left(\underset{r \in Q}{\vee} B_y(r - s) \wedge B_z(t - r) \right) =$$

$$= \underset{s \in Q}{\vee} B_x(s) \wedge \left(\underset{r \in Q}{\vee} B_y(r - s) \wedge B_z(t - s - (r - s)) \right).$$

We denote $C_r = \{r - s \colon s \in Q\}$, the C_r is a countable dense set in $\mathbb{R}^1$. Hence, by Lemma 3.2.4, we have

$$\underset{s \in Q}{\vee} B_x(s) \wedge (\underset{r \in Q}{\vee} B_y(r - s) \wedge B_z((t - s) - (r - s))) =$$

$$= \underset{s \in Q}{\vee} \left(B_x(s) \wedge B_{y+z}(r - s) \right) =$$

$$= B_{x+(y+z)}(t).$$

(iii) Calculate:

$$B_{x+y}(t) = \underset{r \in Q}{\vee} x((-\infty, r)) \wedge y((-\infty, s - r)) =$$

$$= \underset{r \in Q}{\vee} \left(z(u^{-1}((-\infty, r))) \wedge z(v^{-1}((-\infty, t - r))) \right) =$$

$$= \underset{r \in Q}{\vee} \left(z(E) \wedge z(F) \right) = \underset{r \in Q}{\vee} z(E \cap F) =$$

$$= z \left(\underset{r \in Q}{\vee} (u^{-1}((-\infty, r)) \wedge v^{-1}((-\infty, t - r))) \right) =$$

$$= z((u + v)^{-1}((-\infty, t))) = (u + v) \circ z((-\infty, t)) =$$

$$= B_{(u+v) \circ z}(t).$$

(iv) Since $B_{l_\alpha}(t - r) = 0_{\mathbb{X}}$ if $(t - \alpha) < r$ and $B_{l_\alpha}(t - r) = 1_{\mathbb{X}}$ otherwise, we have

$$B_{z+l_\alpha}(t) = \underset{r \geq (t-r)}{\vee} (0_{\mathbb{X}} \wedge B_x(r)) \vee \underset{r \leq (t-r)}{\vee} (1_{\mathbb{X}} \wedge B_x(r)) =$$

$$= \underset{r \leq (t-r)}{\vee} B_x(r) = B_x(t - \alpha) = u_\alpha \circ z((-\infty, t)).$$

(v) We have $B_{\beta x}(r) = \beta x((-\infty, t)) = B_x\left(\frac{r}{\beta}\right)$, $B_{\beta y}(t - r) = B_y\left(\frac{t}{\beta} - \frac{r}{\beta}\right)$ for $\beta > 0_{\mathbb{X}}$, hence

$$B_{\beta x + \beta y}(t) = \bigvee_{r \in Q} B_{\beta x}(r) \wedge B_{\beta y}(t - r) =$$

$$= \bigvee_{r \in Q} B_x\left(\frac{r}{\beta}\right) \wedge B_y\left(\frac{t}{\beta} - \frac{r}{\beta}\right) =$$

$$= B_{x+y}\left(\frac{t}{\beta}\right) = B_{\beta(x+y)}(t).$$

Remark 3.2.7 *If $\mathcal{M}$ consists of crisp subsets, $\mathcal{M}$ is a σ-algebra of subsets of $\mathbb{X}$ (more precisely, $\mathcal{M}$ is a set of all characteristic functions of sets from the given σ-algebra of subsets of the universum $\mathbb{X}$), then the sum of fuzzy observables coincides with the point wisely defined sum.*

Indeed, in this case for x and y there are unique mappings $u, v \colon \mathbb{X} \to \mathbb{R}^1$ such that

$$x(E) = u^{-1}(E) \text{ and } y(F) = v^{-1}(F), E, F \in \mathcal{B}(\mathbb{R}^1),$$

and

$$(x + y)(E) = (u + v)^{-1}(E)$$

for any $E \in \mathcal{B}(\mathbb{R}^1)$ (see Theorem 3.2.5).

Remark 3.2.8 *If $\Theta(\mathcal{M})$ is the set of all fuzzy observables of a fuzzy quantum space $(\mathbb{X}, \mathcal{M})$, then $\Theta(\mathcal{M})$ is a real vector space with respect to the sum.*

Remark 3.2.9 *We define the subtraction of fuzzy observables x and y as follows:*

$$x - y = x + (-y),$$

where

$$(-y)(E) = y(\{t \colon -t \in E\}), E \in \mathcal{B}(\mathbb{R}^1).$$

3.3 CONVERGENCES OF FUZZY OBSERVABLES

Various types of convergences of random variables belong among important concepts of the probability theory. Therefore, the notion of a fuzzy observable is an analogy to the notion of a random variable. In this chapter we study the convergences of fuzzy observables on fuzzy quantum space $(\mathbb{X}, \mathcal{M})$.

We introduce selected theorems of the convergence of random variables, known from the classical probability theory, and we verify its validity in the theory of fuzzy sets. We verify the validity of those statements by the method of F-σ-ideals, which is also described here.

For defining different types of convergence and for the proof of limit theorems on fuzzy quantum space $(\mathbb{X}, \mathcal{M})$ we need some properties of fuzzy state m, which is defined in the first section (Definition 1.10). These properties of a fuzzy state were proved by Piasecki [70], in a different manner also by Dvurečenskij [15]. We summarize them in the following theorem.

Theorem 3.3.1 *Any fuzzy state m on fuzzy quantum space $(\mathbb{X}, \mathcal{M})$ possesses the following properties:*

(i) $m(f \wedge (1_{\mathbb{X}} - f)) = 0$, for every $f \in \mathcal{M}$.

(ii) $m(1_{\mathbb{X}} - f) = 1_{\mathbb{X}} - m(f)$, for every $f \in \mathcal{M}$.

(iii) $m(f \vee g) + m(f \wedge g) = m(f) + m(g)$ for every $f \in \mathcal{M}$.

(iv) If $f_n \nearrow f$ $(f_n \searrow f)$, $f_n, f \in \mathcal{M}$ $(n = 1,2,\dots)$, then $\lim\limits_{n \to \infty} m(f_n) = m(f)$.

(v) If $f, g \in \mathcal{M}$, $f \leq g$, then $m(f) \leq m(g)$.

(vi) If $f, g \in \mathcal{M}$, $f \leq g$, then $m(g) = m(f) + m(f' \wedge g)$.

(vii) $m(f) = m(f \wedge e)$ for any $f \in \mathcal{M}$ and $e \in W_1(\mathcal{M})$.

(viii) $m(f) = m(f \vee l)$ for any $f \in \mathcal{M}$ and $l \in W_0(\mathcal{M})$.

(The symbol $W_0(\mathcal{M})$ denotes the system of all W-empty sets from $\mathcal{M}$ and $W_1(\mathcal{M})$ denotes the set of all W-universum).

Due to Theorem 3.3.1 we prove that any fuzzy state on the fuzzy quantum space $(\mathbb{X}, \mathcal{M})$ is sub-σ-additive.

Lemma 3.3.2 *Let m be a fuzzy state on the fuzzy quantum space $(\mathbb{X}, \mathcal{M})$. Then*

$$m\left(\bigvee_{n=1}^{\infty} f_n\right) \leq \sum_{n=1}^{\infty} m(f_n)$$

or any sequence $\{f_n\} \subset \mathcal{M}$.

Proof. By mathematical induction we can show that for every $n \geq 1$ the following inequality holds

$$m\left(\bigvee_{i=1}^{n} f_n\right) \leq \sum_{i=1}^{n} m(f_i). \tag{3.13}$$

According to property (iii) (Theorem 3.3.1), for $n = 2$ the condition (3.13) holds. Suppose that the condition (3.13) is true for n. Then for $k = n + 1$ we calculate:

$$m((f_1 \vee f_2 \vee ...\vee f_n) \vee f_{n+1}) \leq m(f_1 \vee f_2 \vee ...\vee f_n) + m(f_{n+1}) \leq$$

$$\leq m(f_1) + m(f_2) + ... + m(f_{n+1})$$

and according to property (iv) (Theorem 3.3.1) we have

$$m\left(\bigvee_{i=1}^{\infty} f_n\right) = \lim_{n\to\infty} m\left(\bigvee_{i=1}^{n} f_n\right) \leq \lim_{n\to\infty} \sum_{i=1}^{n} m(f_i) = \sum_{i=1}^{\infty} m(f_i).$$

The mean value of a fuzzy observable on a fuzzy quantum space $(\mathbb{X}, \mathcal{M})$ was defined by Riečan [79] as follows.

Let x be a fuzzy observable, and let m be a fuzzy state. If the integral

$$m(x) := \int_{\mathbb{R}^1} t \, dm_x(t)$$

exists, then $m(x)$ is called the mean value of x in m, where

$$m_x : E \to m(x(E)), E \in \mathcal{B}(\mathbb{R}^1),$$

is a probability measure on $\mathcal{B}(\mathbb{R}^1)$. Moreover, if u is a Borel measurable function, then

$$m(u \circ x) := \int_{\mathbb{R}^1} u(t) \, dm_x(t)$$

in that sense that if one side exists, then the second one exists too, and they are equal. Specially, if

$$u(t) = (t - m(x))^2,$$

then

$$D(x) := m((x - m(x))^2)$$

is called the dispersion of a fuzzy observable x in a fuzzy state m.

Now, we present the general method - method of F-σ-ideals which enables us to reformulate and prove many of known limit theorems of classical probability theory for the fuzzy quantum space $(\mathbb{X}, \mathcal{M})$.

Definition 3.3.3 *Let $(\mathbb{X}, \mathcal{M})$ be a fuzzy quantum space. Subset $I \subset \mathcal{M}$ is called F-σ-ideal on a fuzzy quantum space $(\mathbb{X}, \mathcal{M})$ if it has the following properties:*

(i) $f \wedge f' \in I$ for every $f \in \mathcal{M}$;

(ii) if $f \in \mathcal{M}$, $g \in I$ $f \leq g$, then $f \in I$;

(iii) if $\{f_n\}_{n=1}^{\infty} \subset I$, then $\overset{\infty}{\underset{n=1}{\vee}} f_n \in I$;

(iv) if $f \wedge g \in I$ for any $g \in W_1(\mathcal{M}) = \{g; \ g' \in I\}$, then $f \in I$.

Let m is a fuzzy state on a fuzzy quantum space $(\mathbb{X}, \mathcal{M})$. Denote I_m a set of all fuzzy sets $f \in \mathcal{M}$ for which $m(f) = 0$. Set I_m is F-σ-ideal (as defined). Dvurečenskij and Riečan [21] proved that I_m is F-σ-ideal on a fuzzy quantum space $(\mathbb{X}, \mathcal{M})$.

We define a relation $\sim_m$ on a fuzzy quantum space $(\mathbb{X}, \mathcal{M})$ as follows.

Definition 3.3.4 *If $f, g \in \mathcal{M}$. Then f is in the relation $f \sim_m g$ on $\mathcal{M}$ iff*

$$m(f \wedge g') = 0 = m(g \wedge f'). \tag{3.14}$$

Theorem 3.3.5 *Let $\sim_m$ be the relation defined by the relationship (3.14). Then the relation $\sim_m$ is the congruence relation, i.e. it fulfills the following conditions.*

(i) $f \wedge f' \sim_m 0_{\mathbb{X}}$ for every $f \in \mathcal{M}$.

(ii) If $f \sim_m g$ then $f' \sim_m g'$ for every $f, g \in \mathcal{M}$.

(iii) If $f_n \sim_m g_n$ for $n = 1,2,...$, then $\bigvee\limits_{n=1}^{\infty} f_n \sim_m \bigvee\limits_{n=1}^{\infty} g_n$.

(iv) If $f \sim_m g$, then $g \sim_m f$ for every $f, g \in \mathcal{M}$.

(v) If $f \sim_m g$ and $g \sim_m e$, then $f \sim_m e$ for every $f, g, e \in \mathcal{M}$.

Proof. (*i*) $f \wedge f' \sim_m 0_{\mathbb{X}}$ iff $m(f \wedge f') \wedge 1_{\mathbb{X}} = 0$ and $m(f' \vee f) \wedge 0_{\mathbb{X}} = 0$. The correctness of both equalities is evident from the property (*i*) of Theorem 3.3.1.

The property (*ii*) follows from the Definition 3.3.3.

(*iii*) Let $f_n \sim_m g_n$ for $n = 1,2,....$. Then

$$m\left(\bigvee_{n=1}^{\infty} f_n \wedge \left(\bigvee_{n=1}^{\infty} g_n\right)'\right) = m\left(\bigvee_{n=1}^{\infty} f_n \wedge \bigwedge_{n=1}^{\infty} g_n'\right) \leq m\left(\bigvee_{n=1}^{\infty} (f_n \wedge g_n')\right) \leq$$

$$\leq \sum_{n=1}^{\infty} m(f_n \wedge g_n') = 0$$

(*iv*) and (*v*) are evident.

Dvurečenskij and Riečan in [21] proved that the congruence relation $\sim_m$ decomposes the set $\mathcal{M}$ on equivalence classes. Let

$$\overline{f} = \{g \in \mathcal{M} : g \sim_m f\},$$

then

$$\mathcal{M}/\sim_m = \{\overline{f} : f \in \mathcal{M}\} \tag{3.15}$$

is a Boolean σ-algebra of the sense of Sikorski [97] with the following Boolean operations, which are proven in [21]:

(i) $\bigvee\limits_i \overline{f_i} := \overline{\bigvee\limits_i f_i};$ **(3.16)**

(ii) $\bigwedge\limits_i \overline{f_i} := \overline{\bigwedge\limits_i f_i};$ **(3.17)**

(iii) $\left(\overline{f}\right)' := \overline{f'};$

(iv) $$\overline{f \wedge f'} = \overline{0}_{\mathbb{X}} \neq \overline{1}_{\mathbb{X}} = \overline{f \vee f'}, f \in \mathcal{M}. \tag{3.18}$$

Let $h\colon \mathcal{M} \to \mathcal{M}/\!\sim_m$ be the mapping defined by

$$h(f) = \overline{f}, f \in \mathcal{M}. \tag{3.19}$$

Then according to the properties (3.16) - (3.18) we can say, that the mapping h is a σ-homomorphism from $\mathcal{M}$ on to $\mathcal{M}/\!\sim_m$, i.e.

(i). $h(f \wedge f') = \overline{0}_{\mathbb{X}}$ for every $f \in \mathcal{M}$;

(ii) $h(f') = h(f)'$, $f \in \mathcal{M}$;

(iii) $h\left(\overset{\infty}{\underset{i=1}{\vee}} f_i\right) = \overset{\infty}{\underset{i=1}{\vee}} h(f_i)$, $\{f_i\} \subset \mathcal{M}$.

Theorem 3.3.6 *Let μ be a mapping from a Boolean σ-algebra $\mathcal{M}/\!\sim_m$ into the interval $[0,1]$, defined by*

$$\mu(\overline{f}) = m(f) \tag{3.20}$$

for every $\overline{f} \in \mathcal{M}/\!\sim_m$. Then μ is well defined and satisfies the following conditions:

(i) $0 \leq \mu(\overline{f}) \leq 1$ for every $\overline{f} \in \mathcal{M}/\!\sim_m$;

(ii) $\mu(\overline{1}_{\mathbb{X}}) = 1$;

(iii) $\mu\left(\overset{\infty}{\underset{i=1}{\vee}} \overline{f}_i\right) = \sum_{i=1}^{\infty} \mu(\overline{f}_i)$ if $\overline{f}_i \leq \overline{f}'_j$ for $i \neq j$

i.e. μ is a probability measure on the Boolean σ-algebra $\mathcal{M}/\!\sim_m$.

Proof. If $\overline{f} = \overline{g}$, then $m(f) = m(f \wedge (g \vee g')) = m(f \wedge g) = m(g)$, hence the definition is uniqueness.

(i) Holds due to relationship (3.17).

(ii) $\mu(\overline{1}_{\mathbb{X}}) = m(1_{\mathbb{X}}) = 1$.

(iii) According to the property (iii) of the Theorem 3.3.1 there holds:

$$\mu\left(\bigvee_{i=1}^{\infty} \overline{f}_i\right) = m\left(\bigvee_{i=1}^{\infty} f_i\right) = \lim_{n\to\infty} m\left(\bigvee_{i=1}^{n} f_i\right) = \lim_{n\to\infty} \mu\left(\bigvee_{i=1}^{n} \overline{f}_i\right) =$$

$$= \lim_{n\to\infty} \mu\left(\bigvee_{i=1}^{n} \left(\overline{f}_i \wedge \left(\bigvee_{j=1}^{i-1} \overline{f}_j\right)'\right)\right) =$$

$$= \lim_{n\to\infty} m\left(\bigvee_{i=1}^{n} \left(f_i \wedge \left(\bigvee_{j=1}^{i-1} f_j\right)'\right)\right) =$$

$$= \lim_{n\to\infty} \Sigma_{i=1}^{n}\, m\left(\bigvee_{i=1}^{n} f_i \wedge \left(\bigvee_{j=1}^{i-1} f_j\right)'\right) = \lim_{n\to\infty} \Sigma_{i=1}^{n}\, \mu\left(\overline{f}_i \wedge \bigwedge_{j=1}^{i-1} \overline{f}'_j\right) =$$

$$= \lim_{n\to\infty} \Sigma_{i=1}^{n}\, \mu(\overline{f}_i) = \Sigma_{i=1}^{\infty}\, \mu(\overline{f}_i).$$

Remark 3.3.7 *If x is a fuzzy observable of a fuzzy quantum space $(\mathbb{X}, \mathcal{M})$, then $\overline{x} = h \circ x$ is an observable of a Boolean σ-algebra $\mathcal{M}/\!\sim_m$, therefore $h \circ x$ is a mapping from $\mathcal{B}(\mathbb{R}^1)$ into $(\mathbb{X}, \mathcal{M})$ such that the following conditions hold:*

(i)
$$h \circ x(\emptyset) = \overline{0}_{\mathbb{X}}; \tag{3.21}$$

(ii)
$$h \circ x(E^c) = \overline{h \circ x(E)}', E \in \mathcal{B}(\mathbb{R}^1); \tag{3.22}$$

(iii)
$$h \circ x(\cup_{n=1}^{\infty} E_n) = \bigvee_{n=1}^{\infty} h \circ x(E_n), \{E_n\} \subset \mathcal{B}(\mathbb{R}^1). \tag{3.23}$$

Gudder and Millikin [27] introduced many types of convergences for observables in quantum logics. Motivating by their definition we introduce the following notions.

Definition 3.3.8 *We say that a sequence $\{x_n\}_{n=1}^{\infty}$ of fuzzy observables on the fuzzy quantum space $(\mathbb{X}, \mathcal{M})$ converges to a fuzzy observable x*

(i) in a fuzzy state m, if for every $\varepsilon > 0$, we have

$$\lim_{n\to\infty} m((x_n - x)([-\varepsilon, \varepsilon])) = 1;$$

(ii) almost everywhere in a fuzzy state m, if for every $\varepsilon > 0$, we have

$$m \left(\bigvee_{k=1}^{\infty} \bigwedge_{n=k}^{\infty} ((x_n - x)([-\varepsilon, \varepsilon])) \right) = 1;$$

(iii) everywhere, if

$$\bigwedge_{p=1}^{\infty} \bigvee_{k=1}^{\infty} \bigwedge_{n=k}^{\infty} \left((x_n - x) \left(\left[-\frac{1}{p}, \frac{1}{p} \right] \right) \right) = 1_{\mathbb{X}};$$

(iv) in a mean p, where $1 \leq p < \infty$, if

$$\lim_{n \to \infty} (m|\overline{x}_n - \overline{x}|^p) = \overline{0};$$

(v) everywhere on f, if

$$f \leq \bigwedge_{p=1}^{\infty} \bigvee_{k=1}^{\infty} \bigwedge_{n=k}^{\infty} \left((x_n - x) \left(\left[-\frac{1}{p}, \frac{1}{p} \right] \right) \right);$$

(iv) uniformly on $\overline{f} \in \mathcal{M}$, if for every $\varepsilon > 0$, there is an integer n_0 such that for all $n \geq n_0$:

$$(x_n - x)([-\varepsilon, \varepsilon]) \geq \overline{f};$$

(vii) uniformly, if for every $\varepsilon > 0$ there is an integer n_0 such that for all $n \geq n_0$:

$$(x_n - x)([-\varepsilon, \varepsilon]) = \overline{1};$$

(viii) almost uniformly in a fuzzy state m, if for every $\varepsilon > 0$ there is a fuzzy set $f \in \mathcal{M}$ such that $m(f') \leq \varepsilon$ and a sequence $\{x_n\}_{n=1}^{\infty}$ converges uniformly to x on f.

We have created a Boolean σ-algebra $\mathcal{M}/\sim_m$, whose observables are compatible, *i.e.* their range belongs to the Boolean σ-algebra and we defined on the Boolean σ-algebra $\mathcal{M}/\sim_m$ a measure μ by $\mu(\overline{f}) = m(f)$.

According to the Theorem of Loomis-Sikorski [97], there is a measurable space $(\Omega, \mathcal{S})$ and a σ-homomorphism φ from $\mathcal{S}$ onto $\mathcal{M}/\sim_m$ and due to Varadarajan [103] there are functions $u_1, u_2, \ldots : \Omega \to \mathbb{R}^1$ such that

$$\varphi(u_i^{-1}(E)) = h \circ x_i(E), i = 1, 2, \ldots, E \in \mathcal{B}(\mathbb{R}^1), \tag{3.24}$$

$$\varphi(u^{-1}(E)) = h \circ x(E), \tag{3.25}$$

where $h \circ x$ is an observable of the Boolean σ-algebra $\mathcal{M}/\!\sim_m$.

Moreover, a mapping $\mu_\varphi \colon S \to [0,1]$, defined as (3.19)

$$\mu_\varphi(\Lambda) = \mu(\varphi(\Lambda)), \Lambda \in S,$$

is a probability measure on S.

On the Fig. (**2**) it is illustrated the basic idea of the method of F-σ-ideals.

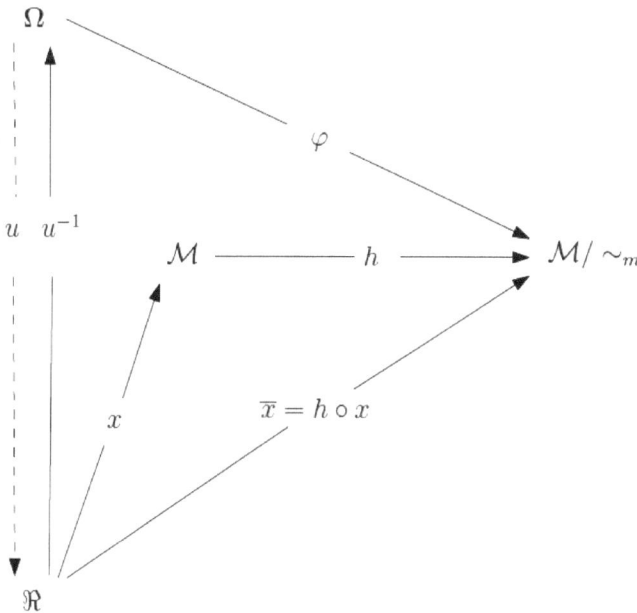

Source: [100]
Fig. (2). The basic idea of the method of F-σ-ideals. Symbol $\Re$ denote space $(\mathbb{R}^1, \mathcal{B}(\mathbb{R}^1))$ and symbol Ω denote space (Ω, S).

We recall that on every Boolean σ-algebra there is always a functional calculus for fuzzy observables, *i.e.* we can produce sums, products and so on (see Varadarajan [103]). Therefore, the next assertion applies.

Lemma 3.3.9 *Let x and y be any two fuzzy observables and h be the homomorphism from $\mathcal{M}$ onto $\mathcal{M}/\!\sim_m$, defined by the relationship 3.19. Then*

(i) $h \circ (x + y) = h \circ x + h \circ y$,

(ii) $h \circ (u \circ x) = u \circ (h \circ x)$

for every Borel function

$$u: \Omega \to \mathbb{R}^1, \text{where } u \circ x: \mathcal{B}(\mathbb{R}^1) \to \mathcal{M},$$

such that

$$u \circ x(E) = x(u^{-1}(E))$$

for every $E \in \mathcal{B}(\mathbb{R}^1)$.

Proof.

    (i) Let $t \in \mathbb{R}^1$, then

$$h \circ (x + y)((-\infty, t)) = h \circ \left(\bigvee_{r \in Q} ((-\infty, r)) \wedge ((-\infty, t - r)) \right) =$$

$$= \bigvee_{r \in Q} h \circ ((-\infty, r)) \vee ((-\infty, t - r)) =$$

$$= \bigvee_{r \in Q} \varphi(u^{-1}((-\infty, r)) \vee (v^{-1}((-\infty, t - r)))) =$$

$$= (\varphi \circ u^{-1} + \varphi \circ v^{-1})((-\infty, t)) =$$

$$= (h \circ x + h \circ y)((-\infty, t)).$$

    (ii) Calculate

$$h \circ (u \circ x)(E) = h \circ \left(x(u^{-1}(E)) \right) = (h \circ x)(u^{-1}(E)) = u \circ (h \circ x),$$

for $E \in \mathcal{B}(\mathbb{R}^1)$

Lemma 3.3.10 *Let m be a fuzzy state of fuzzy quantum space $(\mathbb{X}, \mathcal{M})$ and let x and y be two observables of $(\mathbb{X}, \mathcal{M})$. If $m(x)$ and $m(y)$ exist and are finite, then*

$$m(x + y) = m(x) + m(y).$$

Proof. According to (4.19), there are functions u and v from Ω into $\mathbb{R}^1$ such that

$$\varphi \circ u^{-1}(E) = h \circ x(E),$$

$$\varphi \circ v^{-1}(E) = h \circ y(E), E \in \mathcal{B}(\mathbb{R}^1).$$

If μ_φ is the probability measure on S such that $\mu_\varphi(\Lambda) = m(\varphi(\Lambda))$ for $\Lambda \in S$, then it can be proved that

$$m(x(E)) = \mu_\varphi(u^{-1}(E)),$$

$$m(y(E)) = \mu_\varphi(v^{-1}(E)), E \in \mathcal{B}(\mathbb{R}^1).$$

Hence

$$m(x) = \int_{\mathbb{R}^1} t \, dm_x(t) = \int_\Omega u(\omega) \, d\mu_\varphi(\omega),$$

$$m(y) = \int_{\mathbb{R}^1} t \, dm_y(t) = \int_\Omega v(\omega) \, d\mu_\varphi(\omega).$$

Calculate

$$m((x+y)((-\infty,t))) = m\left(\bigcup_{r \in Q} (x((-\infty,r)) \wedge y((-\infty, t-r)))\right) =$$

$$= \mu\left(\bigcup_{r \in Q} \left(\varphi(u^{-1}(-\infty,r)) \wedge \varphi(v^{-1}(-\infty, t-r))\right)\right) =$$

$$= \mu\left(\varphi((u+v)^{-1}(-\infty,t))\right) =$$

$$= \mu_\varphi((u+v)^{-1}(-\infty,t)).$$

Hence

$$m((x+y)) = \int_\Omega \left((u(\omega) + v(\omega))\right) d\mu_\varphi(\omega) =$$

$$= \int_\Omega u(\omega) \, d\mu_\varphi(\omega) + \int_\Omega v(\omega) \, d\mu_\varphi(\omega) =$$

$$= m(x) + m(y).$$

Theorem 3.3.11 *Let m be a fuzzy state of a fuzzy quantum space $(\mathbb{X}, \mathcal{M})$, $x, x_1, x_2, \ldots$ be fuzzy observables of $(\mathbb{X}, \mathcal{M})$ and $u, u_1, u_2, \ldots$ be functions with properties (3.24) - (3.25). Then*

A) the sequence of fuzzy observables $\{x_n\}_{n=1}^\infty$ converges to a fuzzy observable x

(i) in the fuzzy state m if and only if the sequence of functions $\{u_n\}_{n=1}^\infty$ converges to u in the measure μ_φ;

(ii) almost uniformly in the fuzzy state m *if and only if the sequence of functions* $\{u_n\}_{n=1}^{\infty}$ *converges almost uniformly to* u *in the measure* μ_{φ};

(iii) almost everywhere in the fuzzy state m *if and only if the sequence of functions* $\{u_n\}_{n=1}^{\infty}$ *converges almost everywhere to* u *in the measure* μ_{φ};

(iv) in mean p ($1 \leq p \leq \infty$) *if and only if* $\{u_n\}_{n=1}^{\infty}$ *converges to* u *in mean* p *in the measure* μ_{φ}.

B) If the sequence of fuzzy observables $\{x_n\}_{n=1}^{\infty}$ *converges to the fuzzy observable* x

(v) everywhere, then there is a $\Lambda \in \mathcal{S}$, *such that* $\varphi(\Lambda) = \overline{1}_{\mathbb{X}}$ *and the* $\{u_n\}_{n=1}^{\infty}$ *converges to* u *everywhere on* Λ;

(vi) uniformly, then there is a $\Lambda \in \mathcal{S}$, *such that* $\varphi(\Lambda) = \overline{1}_{\mathbb{X}}$ *and the sequence* $\{u_n\}_{n=1}^{\infty}$ *converges to* u *uniformly on* Λ;

(vii) uniformly on $f \in \mathcal{M}$, *then there is a* $\Lambda \in \mathcal{S}$, *such that* $\varphi(\Lambda) \geq \overline{f}$ *and the sequence* $\{u_n\}_{n=1}^{\infty}$ *converges to* u *uniformly on* Λ;

Conversely, if the sequence of functions $\{u_n\}_{n=1}^{\infty}$, *defined by (4.24) - (4.25) converges to* u

(viii) everywhere, then $\{x_n\}_{n=1}^{\infty}$ *converges to the fuzzy observable* x *everywhere on* $f \in \mathcal{M}$, *where* $\overline{f} = \overline{1}_{\mathbb{X}}$;

(ix) uniformly, then $\{x_n\}_{n=1}^{\infty}$ *converges to the fuzzy observable* x *uniformly on* $f \in \mathcal{M}$, *where* $\overline{f} = \overline{1}_{\mathbb{X}}$;

(x) uniformly on Λ, $\Lambda \in \mathcal{S}$, *then* $\{x_n\}_{n=1}^{\infty}$ *converges to the fuzzy observable* x *uniformly on* $f \in \mathcal{M}$, *where* $\overline{f} = \varphi(\Lambda)$.

Proof.

(i) According to Definition 3.8, a sequence of fuzzy observables $\{x_n\}_{n=1}^{\infty}$ converges to a fuzzy observable x in the fuzzy state m, if for every $\varepsilon > 0$ there holds:

$$\lim_{n \to \infty} m((x_n - x)([-\varepsilon, \varepsilon])) = 1.$$

Calculate:

$$\lim_{n\to\infty} m((x_n - x)([-\varepsilon, \varepsilon])) = \lim_{n\to\infty} \mu(h(x_n - x)([-\varepsilon, \varepsilon])) =$$

$$= \lim_{n\to\infty} \mu((h \circ x_n - h \circ x)([-\varepsilon, \varepsilon])) =$$

$$= \lim_{n\to\infty} \mu(\varphi \circ (u_n - u)^{-1}([-\varepsilon, \varepsilon])) =$$

$$= \lim_{n\to\infty} \mu_\varphi (u_n - u)^{-1}([-\varepsilon, \varepsilon]),$$

and this implies, that the sequence of functions $\{u_n\}_{n=1}^{\infty}$ converges to u in the measure μ_φ and conversely.

(iii) - (iv) we can prove analogously as the property (i).

(v) By the Definition 3.8 a sequence of fuzzy observables $\{x_n\}_{n=1}^{\infty}$ converges to a fuzzy observable x everywhere, if there holds:

$$\bigwedge_{p=1}^{\infty} \bigvee_{k=1}^{\infty} \bigwedge_{n=k}^{\infty} \left((x_n - x)\left(\left[-\frac{1}{p}, \frac{1}{p}\right]\right) \right) = 1_{\mathbb{X}};$$

Calculate:

$$\varphi(\Omega) = h(1_{\mathbb{X}}) = h\left(\bigwedge_{p=1}^{\infty} \bigvee_{k=1}^{\infty} \bigwedge_{n=k}^{\infty} \left((x_n - x)\left(\left[-\frac{1}{p}, \frac{1}{p}\right]\right) \right) \right) =$$

$$= \bigwedge_{p=1}^{\infty} \bigvee_{k=1}^{\infty} \bigwedge_{n=k}^{\infty} h(x_n - x)\left(\left[-\frac{1}{p}, \frac{1}{p}\right]\right) =$$

$$= \bigwedge_{p=1}^{\infty} \bigvee_{k=1}^{\infty} \bigwedge_{n=k}^{\infty} \varphi \circ (u_n - u)^{-1}\left(\left[-\frac{1}{p}, \frac{1}{p}\right]\right) =$$

$$= \varphi\left(\bigcap_{p=1}^{\infty} \bigcup_{k=1}^{\infty} \bigcap_{n=k}^{\infty} (u_n - u)^{-1}\left(\left[-\frac{1}{p}, \frac{1}{p}\right]\right) \right).$$

We denote

$$\Lambda = \left(\bigcap_{p=1}^{\infty} \bigcup_{k=1}^{\infty} \bigcap_{n=k}^{\infty} (u_n - u)^{-1}\left(\left[-\frac{1}{p}, \frac{1}{p}\right]\right) \right),$$

then

$$\varphi(\Lambda) = h(1_{\mathbb{X}}) = \overline{1},$$

which that the sequence of functions $\{u_n\}_{n=1}^{\infty}$ converges to u everywhere on Λ.

(*vi*) and (*vii*) can be proved analogously as a property (*v*).

(*viii*) Let the sequence of functions $\{u_n\}_{n=1}^{\infty}$ converges to u everywhere on Λ, $\varphi(\Lambda) = \overline{1}$, then

$$\bigwedge_{p=1}^{\infty} \bigvee_{k=1}^{\infty} \bigwedge_{n=k}^{\infty} (u_n - u)^{-1} \left(\left[-\frac{1}{p}, \frac{1}{p}\right]\right) = \Lambda.$$

Calculate:

$$\overline{1}_{\mathbb{X}} = \varphi\left(\bigcap_{p=1}^{\infty} \bigcup_{k=1}^{\infty} \bigcap_{n=k}^{\infty} (u_n - u)^{-1} \left(\left[-\frac{1}{p}, \frac{1}{p}\right]\right)\right) =$$

$$= \bigwedge_{p=1}^{\infty} \bigvee_{k=1}^{\infty} \bigwedge_{n=k}^{\infty} \varphi \circ (u_n - u)^{-1} \left(\left[-\frac{1}{p}, \frac{1}{p}\right]\right) =$$

$$= \bigwedge_{p=1}^{\infty} \bigvee_{k=1}^{\infty} \bigwedge_{n=k}^{\infty} h(x_n - x) \left(\left[-\frac{1}{p}, \frac{1}{p}\right]\right) =$$

$$= h\left(\bigwedge_{p=1}^{\infty} \bigvee_{k=1}^{\infty} \bigwedge_{n=k}^{\infty} \left((x_n - x) \left(\left[-\frac{1}{p}, \frac{1}{p}\right]\right)\right)\right) =$$

$$= h(1_{\mathbb{X}}).$$

If we denote

$$f = \bigwedge_{p=1}^{\infty} \bigvee_{k=1}^{\infty} \bigwedge_{n=k}^{\infty} \left((x_n - x) \left(\left[-\frac{1}{p}, \frac{1}{p}\right]\right)\right),$$

then the sequence of fuzzy observables $\{x_n\}_{n=1}^{\infty}$ converges to the fuzzy observable x everywhere on f, $h(f) = \overline{1}_{\mathbb{X}}$. Similarly we can proved (*ix*) and (*x*).

CHAPTER 4

Limit Theorems

Abstract: In this chapter we introduce selected limit theorems on fuzzy quantum space, namely Egorov's theorem, Central limit theorem, Weak and strong law of large numbers, and extreme value theorems for fuzzy quantum space. We also study here the Ergodic theory for fuzzy quantum space and Ergodic theorems and Poincaré recurrence theorems for fuzzy quantum dynamical systems, the Hahn-Jordan decomposition and Lebesgue decomposition for fuzzy quantum space.

Keywords: Egorov's theorem, Central limit theorem, Weak law of large numbers, Strong law of large numbers, Fisher-Tippett, Gnedenko theorem, Balkema, de Haan-Pickands theorem, Birkhoff's individual ergodic theorem, The representation theorem, Individual Ergodic Theorem, Poincaré recurrence theorem, Strong Poincaré recurrence theorem, Hahn-Jordan decomposition, Lebesgue decomposition.

4.1 LIMIT THEOREMS FOR FUZZY QUANTUM SPACE

Using the fact that the sum of observables always exists and it is unique, we formulate and prove selected limit theorems for fuzzy quantum space, such as the Egorov's theorem, the central limit theorem and the law of large numbers, using the method of fuzzy σ-ideals.

Egorov's theorem plays a significant role in the classical theory of probability, stating that the convergence of random variables almost everywhere implies convergence almost uniformly.

In the following, we formulate and prove the Egorov's theorem for fuzzy quantum space $(\mathbb{X}, \mathcal{M})$. We perform to prove in two ways:

I. the direct proof of the Egorov's theorem,

II. the proof using the method of fuzzy σ-ideals.

Theorem 4.1.1 (Egorov's theorem) *If a sequence of fuzzy observables $\{x_n\}_{n=1}^{\infty}$ converges to a fuzzy observable x in a fuzzy state m, then this sequence converges almost uniformly to the fuzzy observable x in the fuzzy state m.*

Renáta Bartková, Beloslav Riečan and Anna Tirpáková

Proof. I. For any $\varepsilon > 0$ we must find $f \in \mathcal{M}$ with property $m(f') < \varepsilon$ such that the sequence of fuzzy observables $\{x_n\}_{n=1}^{\infty}$ converges uniformly to fuzzy observable x on f. We know that for any $\varepsilon = \frac{1}{k}$, $k \geq 1$:

$$m\left(\bigvee_{n=1}^{\infty} \bigwedge_{i=n}^{\infty} \left((x_i - x) \left(\left[-\frac{1}{k}, \frac{1}{k} \right] \right) \right) \right) = 1.$$

We denote

$$g_n^k = \bigwedge_{i=n}^{\infty} \left((x_i - x) \left(\left[-\frac{1}{k}, \frac{1}{k} \right] \right) \right).$$

Then $g_n^k \leq g_{n+1}^k$, $m\left(\bigvee_{n=1}^{\infty} g_n^k \right) = 1$ for every k. Hence for every $\frac{\varepsilon}{2^k}$ there exists $n_0 = n_0(k)$ such that for any $n \geq n_0(k)$ the following inequality holds:

$$m(1_{\mathbb{X}} - g_n^k) < \frac{\varepsilon}{2^k}.$$

We denote

$$f' = \bigvee_{k=1}^{\infty} \left(1_{\mathbb{X}} - g_{n_0(k)}^k \right).$$

According to the sub-σ-additivity of fuzzy state m there holds:

$$m(f') = m\left(\bigvee_{k=1}^{\infty} \left(1_{\mathbb{X}} - g_{n_0(k)}^k \right) \right) \leq \sum_{k=1}^{\infty} m\left(1_{\mathbb{X}} - g_{n_0(k)}^k \right) \leq \sum_{k=1}^{\infty} \frac{\varepsilon}{2^k} = \varepsilon.$$

Now we show that the sequence of fuzzy observables $\{x_n\}_{n=1}^{\infty}$ converges uniformly to a fuzzy observable x on f.

For any $\eta > 0$ there is a positive integer k_0 such that $\frac{1}{k} < \eta$ for every $k > k_0$, . If we use property (iv) of Definition 3.3.8, then for k_0 there is positive integer $n_0 = n_0(k)$ such that for any $n \geq n_0$:

$$(x_n - x)([-\eta, \eta]) \geq (x_n - x)\left(\left[-\frac{1}{k_0}, \frac{1}{k_0} \right] \right) \geq$$

$$\geq \bigwedge_{i=n_0(k)}^{\infty} \left((x_i - x) \left(\left[-\frac{1}{k_0}, \frac{1}{k_0} \right] \right) \right) =$$

$$= g_{n_0(k)}^k \geq f.$$

II. Let $\{u_n\}_{n=1}^{\infty}$ be a sequence of functions and let u be the corresponding function with properties 3.24 - 3.25. According to Theorem 3.3.11 sequence of function $\{u_n\}_{n=1}^{\infty}$ converges to function u in measure μ_{φ}. Due to Egorov's theorem the sequence of function converges almost uniformly to function u in measure μ_{φ}. It is (due to Theorem 3.3.11) equivalent to the convergence of sequence of fuzzy observables $\{x_n\}_{n=1}^{\infty}$ almost uniformly to fuzzy observable x in fuzzy state m.

By using property (iii) of Theorem 3.3.11, the following theorems can be proved: Central limit theorem and the Law of large numbers. Due to the fact that these theorems are corollaries of equivalence of almost everywhere convergences of sequences of fuzzy observables $\{x_n\}_{n=1}^{\infty}$ in fuzzy state m and sequences of functions $\{v_n\}_{n=1}^{\infty}$ in measure μ_{φ}, we can use the proof of both these theorems.

In the following, for formulating the Central limit theorem and the Law of large numbers we introduce the notion of the independence of fuzzy observables $\{x_n\}_{n=1}^{\infty}$ in fuzzy state m. Now we define joint fuzzy observable of fuzzy observables.

Definition 4.1.2 *Let $x_1, x_2, \ldots, x_n$, $n \geq 2$, be a finite system of fuzzy observables on fuzzy quantum space $(X, \mathcal{M})$. A joint fuzzy observable of fuzzy observables $x_1, x_2, \ldots, x_n$ is such a σ-homomorphism $T_n : \mathcal{B}(\mathbb{R}^n) \to \mathcal{M}$ that*

(i) $T_n(A^c) = T_n(A)'$ for every $A \in \mathcal{B}(\mathbb{R}^n)$,

(ii) $T_n(\bigcup_{i=1}^{n} A_i) = \bigwedge\limits_{i=1}^{n} T_n(A_i)$, $A_i \in \mathcal{B}(\mathbb{R}^n)$, $i = 1, 2, \ldots, n$,

(iii) $T_n(\pi_i^{-1}(E) = x_i(E))$ for every $i \in \{1, 2, \ldots, n\}$, $E \in \mathcal{B}(\mathbb{R})$,

where $\pi_i : \mathbb{R}^n \to \mathbb{R}$ is the projection into the i-th coordinate.

By Riečan [89] and Riečan, Neubrunn [92] (Theorem 14.6.2) a sufficient condition for the existence of the joint fuzzy observable of fuzzy observables $x_1, x_2, \ldots, x_n$, $n \geq 2$, is meeting the following condition: $x_i(\emptyset) = x_j(\emptyset)$ for every $i, j \in \{1, 2, \ldots, n\}$.

Definition 4.1.3 *Fuzzy observables $x_1, x_2, \ldots, x_n$ on a fuzzy quantum space $(\mathbb{X}, \mathcal{M})$ are independent in a fuzzy state m, if for every $n \geq 2$ there exists the joint fuzzy observable T_n, and*

$$m(T_n(E_1 \times E_2 \times \ldots \times E_n)) = \prod_{i=1}^{n} m(x_i(E_i)), \qquad (4.1)$$

for any $E_i \in \mathcal{B}(\mathbb{R}^1)$, $i = 1,2,\ldots,n$.

According to the assumption of independence of sequence of fuzzy observables $\{x_n\}_n^\infty$ for every $n \geq 2$ there exists the joint fuzzy observable T_n. To each fuzzy observables $x_i: \mathcal{B}(\mathbb{R}) \to \mathcal{M}$, $i = 1,2,\ldots,n$, exists the observable $\overline{x_i} = h \circ x_i: \mathcal{B}(\mathbb{R}) \to \mathcal{M}/\sim_m$ (Remark 3.3.7) and a real function $u_i: \Omega \to \mathbb{R}^1$, such that $\overline{x_i}(E) = \varphi(u_i^{-1}(E))$ (see Fig. **2**). We define function $\Phi_n: \Omega \to \mathbb{R}^n$, such that

$$\Phi_n(\omega) = (u_1(\omega), u_2(\omega), \ldots, u_n(\omega)), \omega \in \Omega.$$

If

$$\overline{T_n} = h \circ T_n: \mathcal{B}(\mathbb{R}^n) \to \mathcal{M}/\sim_m,$$

then

$$\overline{T_n} = \varphi \circ \Phi_n^{-1}.$$

The main idea of the proof can be illustrated by the Fig. (**3**):

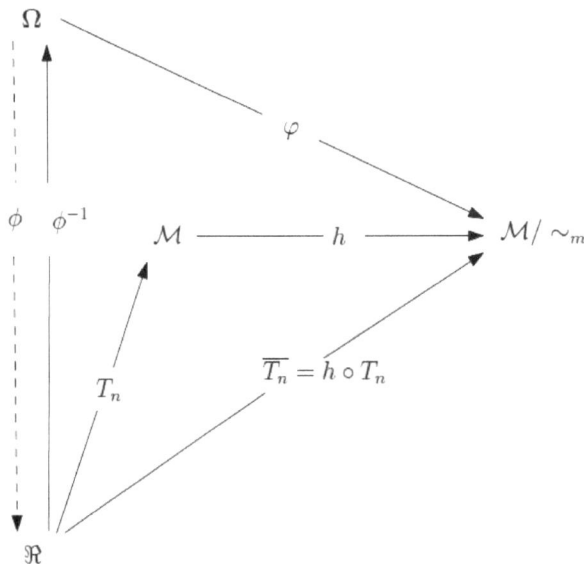

Source: The author of the figure is R. Bartková. The figure has not been published before in any other publication.

Fig. (3). The main idea of the proof of Central limit theorem. Symbol $\mathfrak{R}$ denote space $(\mathbb{R}^1, \mathcal{B}(\mathbb{R}^1))$ and symbol $\mathbf{\Omega}$ denote space $(\Omega, \mathcal{S})$.

Theorem 4.1.4 (Central limit theorem) *Let $\{x_n\}_{n=1}^{\infty}$ be a sequence of independent fuzzy observables, identically distributed in a fuzzy state m, with the mean value a and the variance $\sigma^2 \in (0, \infty)$. Then for any $s \in \mathbb{R}^1$ the following equality holds:*

$$\lim_{n\to\infty} m\left(\frac{1}{\sigma\sqrt{n}}\Sigma_{i=1}^{n}\ (x_i - na)(-\infty, s)\right) = \frac{1}{\sqrt{2\pi}}\int_{-\infty}^{s} e^{-\frac{t^2}{2}}dt.$$

Proof. We define the real function $k_n \colon \mathbb{R}^n \to \mathbb{R}^1$ as follows

$$k_n(r_1, r_2, \ldots, r_n) = \frac{1}{\sigma\sqrt{n}}\Sigma_{i=1}^{n}\ (r_i - na),\ r_i \in \mathbb{R},\ i = 1, 2, \ldots, n.$$

Calculate:

$$m\left(\frac{1}{\sigma\sqrt{n}}\Sigma_{i=1}^{n}\ (x_i - na)(-\infty, s)\right) = m(T_n(k_n^{-1}((-\infty, s)))) =$$

$$= \mu\big(\overline{T_n}(k_n^{-1}((-\infty, s)))\big) =$$

$$= \mu(\varphi \circ \Phi_n^{-1}(k_n^{-1}((-\infty, s)))) =$$

$$= \mu_\varphi(\Phi_n^{-1}(k_n^{-1}((-\infty, s)))) =$$

$$= \mu_\varphi(\{\omega \colon k_n(\Phi_n(\omega)) < s\}),$$

where

$$k_n(\Phi_n(\omega)) = \frac{1}{\sigma\sqrt{n}}\Sigma_{i=1}^{n}\ (u_i - na)$$

Regarding to Theorem 1.7.5, the validity following argument it is obvious.

$$\lim_{n\to\infty} m\left(\frac{1}{\sigma\sqrt{n}}\Sigma_{i=1}^{n}\ (x_i - na)(-\infty, s)\right) = \lim_{n\to\infty}\mu_\varphi(\{\omega \colon k_n(\Phi_n(\omega)) < s\}) =$$

$$= \frac{1}{\sqrt{2\pi}}\int_{-\infty}^{s} e^{\frac{-t^2}{2}}dt.$$

Theorem 4.1.5 (Weak law of large numbers) *Let $\{x_n\}_{n=1}^{\infty}$ be a sequence of independent fuzzy observables, identically distributed in a fuzzy state m, with mean value a. Then*

$$\frac{x_1 + \cdots + x_n}{n} - a$$

converges to the null fuzzy observable o in the fuzzy state m (Definition 3.1.7).

Proof. We define real function $k_n \colon \mathbb{R}^n \to \mathbb{R}^1$ as follows

$$k_n(r_1, r_2, \ldots, r_n) = \tfrac{1}{n}\sum_{i=1}^{n} r_i - a, r_i \in \mathbb{R}, i = 1, 2, \ldots, n.$$

If T_n is the joint fuzzy observable of fuzzy observables $x_1, \ldots, x_n$, then we define fuzzy observable $y_n = T_n \circ k_n^{-1}$.

According to Definition 3.3.8 (*i*), a sequence of fuzzy observables $\{y_n\}_{n=1}^{\infty}$ converges to null fuzzy observable o in fuzzy state m, if for any $\varepsilon > 0$ there holds:

$$\sum_{i=1}^{\infty} m\left(y_n\big((-\varepsilon, \varepsilon)\big)\right) = 1.$$

Calculate:

$$m\left(y_n\big((-\varepsilon, \varepsilon)\big)\right) = m(T_n(k_n^{-1}((-\varepsilon, \varepsilon)))) =$$

$$= \mu\big(\overline{T_n}(k_n^{-1}((-\varepsilon, \varepsilon)))\big) =$$

$$= \mu(\varphi \circ \Phi_n^{-1}(k_n^{-1}((-\varepsilon, \varepsilon)))) =$$

$$= \mu_\varphi(\Phi_n^{-1}(k_n^{-1}((-\varepsilon, \varepsilon)))).$$

If $k_n\big(\Phi_n(\omega)\big) = \tfrac{1}{n}\sum_{i=1}^{n} u_i(\omega) - a$, then according to Theorem 3.3.11 A) (i): the sequence of fuzzy observables $\{y_n\}_{n=1}^{\infty}$ converges to the null fuzzy observable o in the fuzzy state m if and only if the sequence of the functions $\{k_n(\Phi_n)\}_{n=1}^{\infty}$ converges to the null in the measure μ_φ.

According to Theorem 1.10.3 the validity of the arguments is obvious.

Theorem 4.1.6 (Strong law of large numbers) *Let $\{x_n\}_{n=1}^{\infty}$ be a sequence of fuzzy observables independent in a fuzzy state m such that*

$$\sum_{i=1}^{\infty} \frac{1}{i^2} D(x_i) < \infty.$$

Then

$$\frac{1}{n}\sum_{i=1}^{n} (x_i - m(x_i))$$

converges to the null fuzzy observable o almost everywhere in the fuzzy state m.

Proof. We define real function $q_n \colon \mathbb{R}^n \to \mathbb{R}^1$ as follows

$$q_n(r_1, r_2, \dots, r_n) = \frac{1}{n}(r_1 - E(r_1) + r_2 - E(r_2) + \dots + r_n - E(r_n)),$$

where $E(r_i)$ is mean value r_i define as $E(r_i) = \int_{\mathbb{R}^1} t \, d\mu_\varphi(t), r_i \in \mathbb{R}, i = 1, 2, \dots, n$.

Then $q_n\big(\Phi_n(\omega)\big) = \frac{1}{n}\left(\sum_{i=1}^{n}(u_i(\omega) - E(u_i(\omega)))\right)$.

If T_n is a joint fuzzy observable of fuzzy observables $x_1, \dots, x_n$, then we define fuzzy observable $z_n = T_n \circ q_n^{-1} = \frac{1}{n}\left(\sum_{i=1}^{n}(x_i - m(x_i))\right)$.

According to Definition 3.3.8 (*ii*), a seguence of fuzzy observables $\{z_n\}_{n=1}^{\infty}$ converges to null fuzzy observable o almost everywhere in fuzzy state m, if for any $\varepsilon > 0$ there holds:

$$m\left(\bigvee_{k=1}^{\infty}\bigwedge_{n=k}^{\infty}(z_n([-\varepsilon, \varepsilon]))\right) = 1.$$

Then according to Theorem 3.3.11 A) (iii):

sequence fuzzy observables $\{z_n\}_{n=1}^{\infty}$ converges to the null fuzzy observable o almost everywhere in the fuzzy state m if and only if the sequence the function $\{q_n(\Phi_n)\}_{n=1}^{\infty}$ converges to the null almost everywhere in a measure μ_φ.

According to Theorem 1.10.5 the validity of the arguments is obvious.

Extreme Value Theorems for Fuzzy Quantum Space

Let $\{x_n\}_{n=1}^{\infty}$ be a sequence of independent identically distributed fuzzy observables of fuzzy quantum space $(X, \mathcal{M})$. For any $n \geq 1$ we define real function $k_n \colon \mathcal{B}(\mathbb{R}^n) \to \mathbb{R}^1$, as follows

$$k_n(r_1, r_2, \dots, r_n) = \max\{r_1, r_2, \dots, r_n\}.$$

Let T_n be the joint fuzzy observable of fuzzy observables $x_1, x_2, \dots, x_n$. We define the maximum fuzzy observables of $x_1, x_2, \dots, x_n$ as:

$$M_1 = x_1,$$

$$M_n = \max\{x_1, x_2, \dots, x_n\} = T_n \circ k_n^{-1}, n \geq 2,$$

where M_n is the fuzzy observable.

Theorem 4.1.7 (Fisher-Tippett, Gnedenko theorem) *Let $\{x_n\}_{n=1}^{\infty}$ be a sequence of independent fuzzy observables, identically distributed in a fuzzy state m. Let there exist norming constants $a_n > 0$ and $b_n \in \mathbb{R}$ and some non-degenerate distribution function H such that*

$$\lim_{n \to \infty} m\left(\frac{1}{a_n}(M_n - b_n)(-\infty, t)\right) = H(t),$$

for any $t \in \mathbb{R}$. Then H belongs to the type of one of the following three types of standard extreme value distributions:

1. *Gumbel*

2. *Fréchet*

3. *Weibull*

Proof. We define real function $q_n \colon \mathbb{R}^n \to \mathbb{R}^1$ as follows

$$q_n(r_1, r_2, \dots, r_n) = \frac{1}{a_n}(\max\{r_1, r_2, \dots, r_n\} - b_n), r_i \in \mathbb{R}^1, i = 1, 2, \dots, n.$$

Then $q_n(\Phi_n(\omega)) = \frac{1}{a_n}(\max\{u_1(\omega), u_2(\omega), \dots, u_n(\omega)\} - b_n)$ and

$$m\left(\frac{1}{a_n}(M_n - b_n)(-\infty, t)\right) = m(T_n(q_n^{-1}((-\infty, t)))) =$$

$$= \mu\left(\overline{T_n}(q_n^{-1}((-\infty, t)))\right) =$$

$$= \mu(\varphi \circ \Phi_n^{-1}(q_n^{-1}((-\infty, t)))) =$$

$$= \mu_\varphi(\Phi_n^{-1}(q_n^{-1}((-\infty, t)))) =$$

$$= \mu_\varphi(\{\omega \colon q_n(\Phi_n(\omega)) < t\}).$$

We have

$$H(t) = \lim_{n \to \infty} m\left(\frac{1}{a_n}(M_n - b_n)(-\infty, t)\right) = \lim_{n \to \infty} \mu_\varphi(\{\omega \colon q_n(\Phi_n(\omega)) < t\}),$$

then according to Theorem 1.13.11 the validity of the arguments it is obvious.

Now, we define distribution function and excess distribution function on a fuzzy quantum space $(\mathbb{X}, \mathcal{M})$.

Definition 4.1.8 *Let* $m: \mathcal{M} \to [0,1]$ *be a fuzzy state and* $x: \mathcal{B}(\mathbb{R}) \to \mathcal{M}$ *be a fuzzy observable on a fuzzy quantum space* $(\mathbb{X}, \mathcal{M})$. *For any* $t \in \mathbb{R}^1$ *we define function* $F_x: \mathbb{R}^1 \to [0,1]$ *as follow*

$$F_x(t) = m\left(x((-\infty, t))\right)$$

The function F_x is called the distribution function of an observable x on a fuzzy quantum space $(\mathbb{X}, \mathcal{M})$.

Proposition 4.1.9 *If the function* F_x *is the distribution function of an observable* x *on a fuzzy quantum space* $(\mathbb{X}, \mathcal{M})$, *then it satisfies the following conditions*:

(i) F_x *is non-decreasing,*

(ii) F_x *is left continuous,*

(iii) $\lim\limits_{n \to \infty} F_x = 1,$

(iv) $\lim\limits_{n \to -\infty} F_x = 0.$

Proof.

(i) Let $s < t$, $s, t \in \mathbb{R}^1$, then $x((-\infty, s)) \leq x((-\infty, t))$, it follows that

$$F_x(s) = m\left(x((-\infty, s))\right) \leq m\left(x((-\infty, t))\right) = F_x(t).$$

We proved, that function the F_x is non - decreasing.

(ii) Let $t_n \nearrow t$, $t_n, t \in \mathbb{R}^1$, $n = 1, 2, \ldots$, then $x((-\infty, t_n)) \nearrow x((-\infty, t))$, it follows that

$$F_x(t_n) = m\left(x((-\infty, t_n))\right) \nearrow m\left(x((-\infty, t))\right) = F_x(t).$$

We proved, that the function F_x is left continuous.

(iii) Let $t_n \nearrow \infty$, $t_n \in \mathbb{R}^1$, $n = 1, 2, \ldots$, then

$x\big((-\infty, t_n)\big) \nearrow x\big((-\infty, \infty)\big) = 1_{\mathbb{X}}$, it follows that

$$F_x(t_n) = m\left(x\big((-\infty, t_n)\big)\right) \nearrow m\left(x\big((-\infty, \infty)\big)\right) = m(1_{\mathbb{X}}) = 1.$$

We proved, that $\lim\limits_{n \to \infty} F_x = 1$.

(iv) Let $t_n \searrow -\infty$, $t_n \in \mathbb{R}$, $n = 1, 2, \ldots$, then

$$x\big((-\infty, t_n)\big) \searrow x\big((-\infty, -\infty)\big) = 0_{\mathbb{X}}, \text{ it follows that}$$

$$F_x(t_n) = m\left(x\big((-\infty, t_n)\big)\right) \searrow m\left(x\big((-\infty, -\infty)\big)\right) = m(0_{\mathbb{X}}) = 0.$$

We proved, that $\lim\limits_{n \to -\infty} F_x = 0$.

Now we define the excess distribution function $\tilde{F}_w$ on a fuzzy quantum space $(\mathbb{X}, \mathcal{M})$.

Definition 4.1.10 *For $w > 0$ we define excess distribution function $\tilde{F}_w$ on a fuzzy quantum space $(\mathbb{X}, \mathcal{M})$ as follows*

$$\tilde{F}_w(t) = \frac{\tilde{F}(t+w) - \tilde{F}(w)}{1 - \tilde{F}(w)},$$

for every $0 < t < \omega(\tilde{F})$, where $\omega(\tilde{F}) = \sup\{t; \tilde{F}(t) < 1\}$. Value $\omega(\tilde{F})$ will be called the right endpoint of the distribution function $\tilde{F}$.

Theorem 4.1.11 (Balkema, de Haan-Pickands) *For sufficiently large w, the excess distribution $\tilde{F}_w$ converges to the GPD. Parameter $\beta = \beta(w)$ is depending on threshold w, and for every $\alpha > 0$*

$$\lim_{w \to \omega(F_x)} \sup_{0 \le t \le \omega(F_x) - w} \left| \tilde{F}_w(t) - G_{\alpha, \beta(w)}(t) \right| = 0.$$

Proof. Let $x_i \colon \mathcal{B}(\mathbb{R}^1) \to M$, $i = 1, 2, \ldots, n$ be fuzzy observables. Then there exist observables $\overline{x_i} = h \circ x_i \colon \mathcal{B}(\mathbb{R}^1) \to M / I_m$ and real functions $u_i \colon \Omega \to \mathbb{R}^1$, such that $\overline{x_i}(E) = \varphi(u_i^{-1}(E))$. Then

$$\tilde{F}(t) = m\left(x\big((-\infty, t)\big)\right) = \mu\left(\bar{x}_i\big((-\infty, t)\big)\right) =$$

$$\mu\left(\varphi \circ u_n^{-1}\left((-\infty, t)\right)\right) = \mu_\varphi\left(u_n^{-1}\left((-\infty, t)\right)\right) = \mu_\varphi(\{\omega: u_n(\omega) < t\}) = F(t),$$

(where $t \in \mathbb{R}$) is the distribution function of real random variable u. It is obvious

$$\tilde{F}_w(t) = \frac{\tilde{F}(t+w) - \tilde{F}(w)}{1 - \tilde{F}(w)} = \frac{F(t+w) - F(w)}{1 - F(w)} = F_w(t).$$

Regarding to Theorem 1.14.2, the proof is done.

4.2 ERGODIC THEORY ON FUZZY QUANTUM SPACE

In this section we study the possibility of extending the validity of some ergodic theorems for fuzzy quantum space $(\mathbb{X}, \mathcal{M})$.

The first authors who were interested in the ergodic theory on fuzzy quantum space, were Harman and Riečan [34]. They proved the validity of the Birkhoff's individual ergodic theorem for the compatible case.

In this section we prove that a more general form of Birkhoff's individual ergodic theorem applies on fuzzy quantum space $(\mathbb{X}, \mathcal{M})$. We also formulate and prove the individual ergodic theorem and the maximal ergodic theorem for fuzzy quantum space $(\mathbb{X}, \mathcal{M})$. Furthermore, we generalize Poincaré recurrence theorems, which were proved for quantum logics by Dvurečenskij [15], and we also generalize some results of Mesiar [56]. Fuzzy analogies of Mesiar's ergodic theorems was presented in [102] and Birkhoff's individual ergodic theorem and the maximal ergodic theorem for the case of fuzzy dynamical systems was proved in [51].

At first we introduce some basic notions for a fuzzy quantum space from the ergodic theory.

Definition 4.2.1 *A mapping* $\tau: \mathcal{M} \to \mathcal{M}$ *such that*

(i). $\tau(f') = (\tau(f))'$, *if* $f \in \mathcal{M}$;

(ii). $\tau\left(\bigvee_{i=1}^{\infty} f_i\right) = \bigvee_{i=1}^{\infty} \tau(f_i)$, *if* $\{f_i\}_{i=1}^{\infty} \subset \mathcal{M}$,

is called a homomorphism of a fuzzy quantum space $(\mathbb{X}, \mathcal{M})$.

Definition 4.2.2 *We say that a homomorphism* τ *of* $(\mathbb{X}, \mathcal{M})$ *is invariant in a fuzzy state m if*

$$m(\tau(f)) = m(f),$$

for every $f \in \mathcal{M}$.

Definition 4.2.3 *A homomorphism τ of $(\mathbb{X}, \mathcal{M})$ invariant in a fuzzy state m is ergodic in m, if the statement*

$$m(f \wedge \tau(f')) = 0 = m(\tau(f) \wedge f')$$

implies $m(f) \in \{0,1\}$.

Remark 4.2.4 If τ is a homomorphism and x is a fuzzy observable of $(\mathbb{X}, \mathcal{M})$, then

$$\tau \circ x : E \to \tau(x(E)), E \in \mathcal{B}(\mathbb{R}^1)$$

is a fuzzy observable of $(\mathbb{X}, \mathcal{M})$, too.

For the proof of Brushoff's individual ergodic theorem on an a fuzzy quantum space $(\mathbb{X}, \mathcal{M})$ we will use the method of a F-σ-ideals as well as the properties of a homomorphism τ which we will formulate in next lemmas.

Lemma 4.2.5 *Let $\overline{\tau}$ be a homomorphism of $(\mathbb{X}, \mathcal{M})$ invariant in a fuzzy state m. The mapping $\overline{\tau}$ from the Boolean σ-algebra $\mathcal{M}/\sim_m$ into $\mathcal{M}/\sim_m$ defined via*

$$\overline{\tau}(\overline{f}) = \overline{\tau(f)}, f \in \mathcal{M},$$

is a homomorphism of the Boolean σ-algebra $\mathcal{M}/\sim_m$ such that the following conditions are satisfied:

(i) $\overline{\tau}(\overline{0}_{\mathbb{X}}) = \overline{0}_{\mathbb{X}}$;

(ii) $\overline{\tau}(\overline{f'}) = (\overline{\tau}(\overline{f}))', f \in \mathcal{M}$;

(iii) $\overline{\tau}\left(\bigvee_{i=1}^{\infty} \overline{f}_i \right) = \bigvee_{i=1}^{\infty} \overline{\tau}(\overline{f}_i), f_i \subset \mathcal{M}$;

moreover $\overline{\tau}$ is invariant in state μ, i.e.

$$\mu(\overline{\tau}(\overline{f})) = \mu(\overline{f}),$$

for every $f \in \mathcal{M}$.

Proof. (i) $\overline{\tau}(\overline{0}_{\mathbb{X}}) = \overline{\tau(0_{\mathbb{X}})} = \overline{0}_{\mathbb{X}}$.

(ii) $\overline{\tau}(\overline{f}') = \overline{\tau}(\overline{f'}) = \overline{\tau(f')} = \overline{\tau(f)'} = \overline{\tau(f)}' = (\overline{\tau}(\overline{f}))'$.

(iii) $\overline{\tau}\left(\bigvee_{i=1}^{\infty} \overline{f}_i\right) = \overline{\tau}\left(\overline{\bigvee_{i=1}^{\infty} f_i}\right) = \overline{\tau\left(\bigvee_{i=1}^{\infty} f_i\right)} = \overline{\bigvee_{i=1}^{\infty} \tau(f_i)} = \bigvee_{i=1}^{\infty} \overline{\tau(f_i)} = \bigvee_{i=1}^{\infty} \overline{\tau}(\overline{f}_i)$.

Remark 4.2.6 *Let x be a fuzzy observable on a fuzzy quantum space $(\mathbb{X}, \mathcal{M})$. Then $y = h \circ x$ is an observable on the Boolean σ-algebra $\mathcal{M}/\sim_m$, i.e. the next properties hold:*

(i) $y(\emptyset) = \overline{0}_{\mathbb{X}}$;

(ii) $y(E^c) = y(E)', E \in \mathcal{B}(\mathbb{R}^1)$;

(iii) $y(\bigcup_{i=1}^{\infty} E_i) = \bigvee_{i=1}^{\infty} y(E_i), \{E_i\} \subset B(\mathbb{R}^1)$.

Lemma 4.2.7 *If τ is invariant in a fuzzy state m, then for any $n = 1,2,\ldots$, we have:*

(i) $\overline{\tau}^n \circ y = h \circ \tau^n \circ x$.

(ii) Let $\mathcal{A}$ be the minimal Boolean sub-σ-algebra of $\mathcal{M}/\sim_m$ containing all rangers of $\tau^n \circ y$, $n = 1,2,\ldots,$. Then $\overline{\tau}(\overline{f}) \in \mathcal{A}$ for any $\overline{f} \in \mathcal{A}$.

Proof. (i) It is evident.

(ii) We indicate $\mathcal{A}_0 = \{\overline{f} \in \mathcal{A} : \overline{\tau}(\overline{f}) \in \mathcal{A}\}$. Then $\overline{0}_{\mathbb{X}}, \overline{1}_{\mathbb{X}} \in \mathcal{A}_0$, and $\mathcal{A}_0$ is the Boolean sub-σ-algebra of $\mathcal{M}/\sim_m$ containing all $\overline{\tau}^n \circ y, n = 1,2,\ldots$. Then $\mathcal{A}_0 = \mathcal{A}$.

The following picture shows the basic idea of using of the method of F-σ- ideals for verification of extending possibility of ergodic theory on the fuzzy quantum space $(\mathbb{X}, \mathcal{M})$.

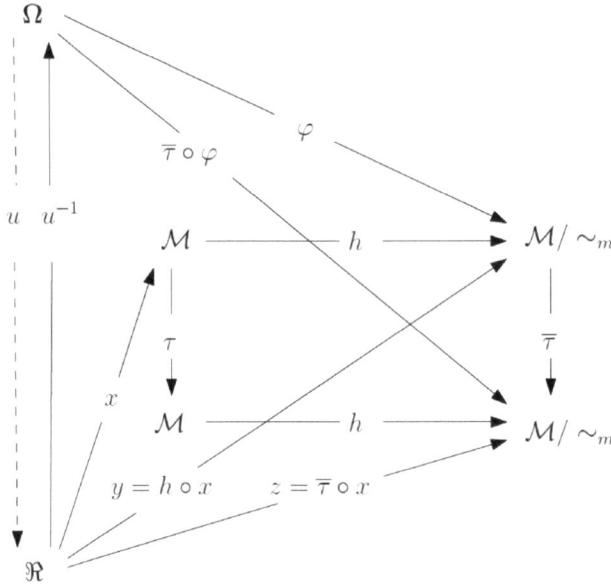

Source: [100]

Fig. (4). The basic idea of the method of F-σ-ideals for verification of validity of ergodic theorems on an F-quantum space. Symbol $\mathfrak{R}$ denote space $(\mathbb{R}^1, \mathcal{B}(\mathbb{R}^1))$ and symbol Ω denote space $(\Omega, \mathcal{S})$.

Theorem 4.2.8 (Birkhoff's individual ergodic theorem) *Let x be a fuzzy observable on a fuzzy quantum space $(\mathbb{X}, \mathcal{M})$ and τ be a homomorphism of the fuzzy quantum space $(\mathbb{X}, \mathcal{M})$ ergodic in a fuzzy state m. Suppose $m(x) = 0$. Then*

$$\frac{1}{n}\sum_{i=1}^{n} \tau^i \circ x \to o \; a.e. \, [m]. \tag{4.2}$$

Proof. It is evident that the Boolean sub-σ-algebra $\mathcal{A}$ in Lemma 4.2.7 has a countable generator. Due to Varadarajan [103], there is an observable $z: \mathcal{B}(\mathbb{R}^1) \to \mathcal{M}/\!\sim_m$ such that

$$\{z(E): E \in \mathcal{B}(\mathbb{R}^1)\} = \mathcal{A},$$

and moreover, there is a sequence of real-valued Borel functions $\{u_n\}_{n=1}^{\infty}$, such that

$$\left(\overline{\tau}^n \circ y\right)(E) = z(u_n^{-1}(E)), E \in \mathcal{B}(\mathbb{R}^1), n = 0,1,\ldots$$

and u_n is essentially unique in the sense that if

$$z(u_n^{-1}(E)) = z(v_n^{-1}(E)) \; E \in \mathcal{B}(\mathbb{R}^1),$$

then

$$z(\{t: u_n(t) \neq v_n(t)\}) = 0_{\mathbb{X}}.$$

From the construction of z it follows that $\overline{\tau}$ is z-measurable, hence

$$\overline{\tau}\big(z(\mathcal{B}(\mathbb{R}^1))\big) \le z(\mathcal{B}(\mathbb{R}^1)).$$

Due to Dvurečenskij and Riečan [21] it is possible iff there is a Borel measurable transformation $T: \mathbb{R}^1 \to \mathbb{R}^1$ such that

$$\overline{\tau}(z(E)) = z(T^{-1}(E)), E \in \mathcal{B}(\mathbb{R}^1).$$

Therefore

$$\overline{\tau^n}(z(E)) = z(T^{-n}(E)), E \in \mathcal{B}(\mathbb{R}^1).$$

and

$$\overline{\tau}^n \circ y(E) = \overline{\tau}^n\big(z(u^{-1}(E))\big) = z((u \circ T^{-n})^{-1}(E)) = z(u_n^{-1}(E)).$$

Due to Varadarajan [103], we may assume without loss of generality that $u_n = u \circ T^{-n}$, $n = 1,2,...$, for some Borel function u.

For observables in a Boolean σ-algebra $\mathcal{M}/\sim_m$, there is a well-known [103] way of definition of their sum, and convergence almost everywhere of observables in $\mathcal{M}/\sim_m$ is same that for fuzzy observables. Therefore

$$\frac{1}{n}\sum_{i=1}^n \tau^n \circ x \to o \ [m],$$

iff

$$\frac{1}{n}\sum_{i=1}^n \overline{\tau}^n \circ y \to \overline{o} \ a.e. \ [\mu],$$

where $\overline{o}(E) = \overline{0}_{\mathbb{X}}$ if $0 \notin E$ and $\overline{o}(E) = \overline{1}_{\mathbb{X}}$ in other cases. The latest convergence is true iff

$$\frac{1}{n}\sum_{i=1}^n u(T^i) \circ z \to \overline{o} \ a.e. \ [\mu],$$

which is possible iff

$$\frac{1}{n}\sum_{i=1}^n u(T^i(t)) \circ z \to 0 \ a.e. \ [\mu_z],$$

where $\mu_z: \mathcal{B}(\mathbb{R}^1) \to [0,1]$ is a probability measure on $\mathcal{B}(\mathbb{R}^1)$ defined by

$$\mu_z(E) = \mu(z(E)), E \in \mathcal{B}(\mathbb{R}^1).$$

On the other hand,

$$m(x) = \int_{\mathbb{R}^1} t \, dm_x(t) = \int_{\mathbb{R}^1} t \, d\mu_z(t) = \int_{\mathbb{R}^1} u(t) \, d\mu_z(t) = 0_{\mathbb{X}}.$$

Take into account a dynamic system $(\mathbb{R}^1, \mathcal{B}(\mathbb{R}^1), \mu_z, T)$. Then T is μ_z-invariant and ergodic in $[\mu_z]$, *i.e.*

(i) $\mu_z(T^{-1}(E)) = \mu_z(E)$, for every $E \in \mathcal{B}(\mathbb{R}^1)$;

(ii) $T^{-1}(E) = E$ implies $\mu_z(E) \in \{0,1\}$.

Therefore, due to Halmos [32], the Birkhoff's individual ergodic theorem holds for u and consequently (4.2) is proved.

Now, we shall prove a more general of individual ergodic theorems on a fuzzy quantum space $(\mathbb{X}, \mathcal{M})$ without a presumption of ergodicity by using the method of the representation of fuzzy observables (Dvurečenskij [11]). For the formulation and the proof we need the following notions.

Lemma 4.2.9 *(Piasecki [67]) Let $(\mathbb{X}, \mathcal{M})$ be a fuzzy quantum space. Then the system*

$$K(\mathcal{M}) = \left\{ A \subseteq \mathbb{X} : \text{there exists } f \in \mathcal{M} : \left\{ f > \tfrac{1}{2} \right\} \subseteq A \subseteq \left\{ f \geq \tfrac{1}{2} \right\} \right\}$$

is a σ-algebra in $\mathbb{X}$.

Lemma 4.2.10 *(Piasecki [67]) Let m be a fuzzy state on fuzzy quantum space $(\mathbb{X}, \mathcal{M})$. The mapping P_m from $K(\mathcal{M})$ to $[0,1]$, such that*

$$P_m(A) = m(A),$$

$f \in \mathcal{M}$, $A \subseteq \mathbb{X}$, $\left\{ f > \tfrac{1}{2} \right\} \subseteq A \subseteq \left\{ f \geq \tfrac{1}{2} \right\}$, is a probability on $K(\mathcal{M})$.

The following theorem for a fuzzy quantum space $(\mathbb{X}, \mathcal{M})$ was formulated and proved by Dvurečenskij [11].

Theorem 4.2.11 (The representation theorem) *(Dvurečenskij [11]) Let x be a fuzzy observable on a fuzzy quantum space $(\mathbb{X}, \mathcal{M})$. Then there exists a real-valued $K(\mathcal{M})$-measurable function u from $\mathbb{X}$ into $\mathbb{R}$, such that*

$$\left\{ x(E) > \tfrac{1}{2} \right\} \subseteq u^{-1}(E) \subseteq \left\{ x(E) \geq \tfrac{1}{2} \right\}, E \in \mathcal{B}(\mathbb{R}^1). \tag{4.3}$$

If there exists an other function v such that

$$\left\{x(E) > \frac{1}{2}\right\} \subseteq v^{-1}(E) \subseteq \left\{x(E) \geq \frac{1}{2}\right\}, E \in \mathcal{B}(\mathbb{R}^1).$$

Then

$$\{y \in \mathbb{X} : u(y) \neq v(y)\} \subseteq \left\{x(\emptyset) = \frac{1}{2}\right\}.$$

Conversely, let u be any real-valued $K(\mathcal{M})$-measurable function from $\mathbb{X}$ into $\mathbb{R}$. Then there exists a fuzzy observable with property (4.3), then

$$x(E) \wedge x(E^c) \leq \left(\frac{1}{2}\right)_{\mathbb{X}} \text{ for every } E \in \mathcal{B}(\mathbb{R}^1).$$

In the following part, we introduce the notion fuzzy equality of two fuzzy observables and the notion of a dynamical system. We will use these notions in the proof of the individual ergodic theorem. Kôpka and Chovanec [40] introduced the notion of fuzzy equality of fuzzy observables as follows:

Let x and y be two fuzzy observables on a fuzzy quantum space $(\mathbb{X}, \mathcal{M})$. We say, that x and y are fuzzy equal (write $x \overset{f}{=} y$) if

$$x(E) \wedge y(E^c) \in I_0, x(E^c) \wedge y(E) \in I_0 \text{ for every } E \in \mathcal{B}(\mathbb{R}^1),$$

where

$$I_0 = \left\{f \in \mathcal{M} : \text{there is } e \geq \frac{1}{2}, e \in \mathcal{M}, \text{such that } f \wedge e \leq \frac{1}{2}\right\}.$$

Markechová in [47] defined the dynamical system as follows. The dynamical system is a quadruple $(\mathbb{X}, \mathcal{M}, m, \tau)$ where $(\mathbb{X}, \mathcal{M})$ is a fuzzy quantum space, m is a fuzzy state and τ is a homomorphism from $\mathcal{M}$ into $\mathcal{M}$ which preserves measure.

Theorem 4.2.12 (Individual Ergodic Theorem) *Let $(\mathbb{X}, \mathcal{M}, m, \tau)$ be a dynamical system and x be an integrable fuzzy observable on a fuzzy quantum space $(\mathbb{X}, \mathcal{M})$. Then there is almost everywhere in m limit*

$$\lim_{n \to \infty} \frac{1}{n} \sum_{i=1}^{n} \tau^i \circ x$$

and a fuzzy observable x^* which $x^* \in L_1$, $\tau \circ x^* \overset{f}{=} x^*$, such that

$$\int x^* dm = \int x dm,$$

and

$$\lim_{n \to \infty} \frac{1}{n} \sum_{i=1}^{n} \tau^i \circ x = x^* \ a.e. \ [m].$$

Proof. We will consider the Boolean σ-algebra $\mathcal{M}/\sim_m$, where $I_0 = \{f \in \mathcal{M}:$ there is $e \geq \frac{1}{2}, e \in \mathcal{M}$, such that $f \wedge e \leq \frac{1}{2}\}$. Dvurečenskij and Riečan [21] proved, that I_0 and I_m, where $I_m = \{f \in \mathcal{M}: m(f) = 0\}$ are σ-ideals on a fuzzy quantum space $(\mathbb{X}, \mathcal{M})$, such that I_0 is the smallest F-σ-ideal, *i.e.* $I_0 \subseteq I_m$.

Let $\bar{\tau}$ be a homomorphism on $\mathcal{M}/\sim_m$, such that $\bar{\tau}(\bar{f}) = \overline{\tau(f)}$, where $\bar{f} = \{g \in \mathcal{M}: g \wedge f', g' \wedge f \in I_0\}$. Due to the invariancy of the measure m in τ we get that $\bar{\tau}$ is invariant in μ on $\mathcal{M}/\sim_m$ and it is defined *via*

$$\mu(\bar{f}) = m(f), \bar{f} \in \mathcal{M}/\sim_m.$$

Let $y = h \circ x$, where h is a homomorphism $\mathcal{M}$ into $\mathcal{M}/\sim_m$, such that $h(f) = \bar{f}$. Then

$$\bar{\tau}^n \circ y = \overline{h \circ \tau^n \circ x}, n = 1, 2, \ldots$$

and $\mathcal{A}$ is a Boolean sub-σ-algebra in $\mathcal{M}/\sim_m$, which includes all ranges of $\bar{\tau}^n \circ y$. Due to Varadarajan [103], there is a σ-homomorphism z from $\mathcal{B}(\mathbb{R}^1)$ into $\mathcal{A}$, such that $\mathcal{A} = z(\mathcal{B}(\mathbb{R}^1))$. It is evident, that $\bar{\tau}$ is z-measurable, *i.e.*

$$\bar{\tau} \circ z(\mathcal{B}(\mathbb{R}^1)) \subseteq z(\mathcal{B}(\mathbb{R}^1)),$$

this is possible iff there is, by Dvurečenskij and Riečan [23], a Borel measurable transformation T from $\mathbb{R}^1$ into $\mathbb{R}^1$ such that

$$\bar{\tau}(z(E)) = z(T^{-1}(E)), E \in \mathcal{B}(\mathbb{R}^1), \tag{4.4}$$

and due to Varadarajan [103], we may to find a sequence of functions u_n such that

$$\bar{y}_n = z \circ u_n^{-1}, n = 1, 2, \ldots,$$

and

$$u_n = u \circ T^n.$$

It is evident, that the function $\mu_z: \mathcal{B}(\mathbb{R}^1) \to [0,1]$, defined *via*

$$\mu_z(E) = \mu(z(E)), E \in \mathcal{B}(\mathbb{R}^1)$$

is a probability on $\mathcal{B}(\mathbb{R}^1)$ and $(\mathbb{R}^1, \mathcal{B}(\mathbb{R}^1), \mu_z, T)$ is a dynamical system,

$$\int u \, d\mu_z = m(x).$$

Due to the classical individual ergodic theorem (Neubrunn and Riečan [62], Theorem 12.47), there exists $u^* \colon \mathbb{R}^1 \to \mathbb{R}^1$ such that

$$u^* \circ T = u^* \in L_1(\mu_z),$$

$$\lim_{n \to \infty} \frac{1}{n} \sum_{i=0}^{n-1} u \circ T^i = u^* \text{ a. e. } [\mu_z].$$

We define $z^* = z \circ u^{*-1}$, then

$$\frac{1}{n} \sum_{i=1}^{n} \tau^{-i} \circ z \to z^* \text{ a. e. } [\mu].$$

and for every $E \in \mathcal{B}(\mathbb{R}^1)$ due to (4.4) we have

$$z^*(E) = z \circ u^{*-1}(E) = z \circ (T^{-1} u^{*-1}(E)) = \bar{\tau} \circ z(u^{-1}(E)) = \bar{\tau} \circ z^*(E),$$

i.e.

$$\bar{\tau} \circ z^* = z^*.$$

It is evident that

$$\mu(z^*) = \int u^* d\mu_z.$$

Let us consider the measurable space $(\Omega, K(\mathcal{M}))$. Due to Loomis-Sikorski analogy [11] for a fuzzy quantum space $(\mathbb{X}, \mathcal{M})$, there is a σ-homomorphism φ from $K(\mathcal{M})$ onto $\mathcal{M}/\sim_m$, and due to Varadarajan [103], there is $K(\mathcal{M})$-measurable function $v^*, v_1, v_2, \ldots$ from $\mathbb{X}$ into $\mathbb{R}^1$ such that

$$\varphi(v^{*-1}(E)) = z^*(E),$$

$$\varphi(v_n^{-1}(E)) = y_n(E), n = 1,2,\ldots$$

Therefore

$$\frac{1}{n} \sum_{i=1}^{n} v^* \to v^* \text{ a. e. } [\mu_\varphi],$$

where μ_φ is a probability on $K(\mathcal{M})$, defined *via*

$$\mu_\varphi(A) = \mu(\varphi(A)), A \in K(\mathcal{M}).$$

The representation Theorem (Theorem 4.2.11) implies, that there exists a fuzzy observable x^* such that

$$\left\{x^*(E) > \tfrac{1}{2}\right\} \subseteq v^{*^{-1}}(E) \subseteq \left\{x^*(E) \geq \tfrac{1}{2}\right\}$$

for every $E \in \mathcal{B}(\mathbb{R}^1)$. Moreover $\overline{x^*} = z^*$, therefore the property holds iff

$$\tfrac{1}{n}\sum_{i=1}^{n} \tau^i \circ x = x^* \ a.\,e.\,[m].$$

Calculate

$$m(x^*) = \int_{\mathbb{X}} v^* d\mu_\varphi = \int_{\mathbb{R}^1} u^* d\mu_z = \int d\mu_{\varphi_z} = m(x).$$

Because

$$\overline{\tau} \circ \overline{x}^* = \tau \circ z^* = z^* = \overline{x^*}$$

and, due to the property (iii) of the representation theorem (Theorem 4.2.11) of a representation of fuzzy observables, we get

$$\tau \circ x^*(E) \wedge x^*(E^c) \in I_0,$$

for every $E \in \mathcal{B}(\mathbb{R}^1)$, *i.e.*

$$\tau \circ x^* \stackrel{f}{=} x^*$$

Remark 4.2.13 *The Theorem 4.2.12 (Individual Ergodic Theorem) is a more general form of the individual ergodic theorem without the assumption of an ergodicity. The Theorem 4.2.12 also implies Theorem 4.2.8 (Birkoff's individual ergodic theorem)*

For additional validation of ergodic properties of the fuzzy quantum space it is necessary to introduce the notion of a product of fuzzy observables and, the notion of an indefinite integral. These notions were introduced by Dvurečenskij [10] and they are defined as follows.

Definition 4.2.14 *The product of two fuzzy observables x and y is defined via*

$$x \cdot y = \frac{(x+y)^2 \quad x^2 \quad y^2}{2}.$$

Definition 4.2.15 *Let f be a fuzzy set of $\mathcal{M}$. Then an indefinite integral of fuzzy observable x on fuzzy set $f \in \mathcal{M}$ is the expression*

$$\int_f x dm = \int x \cdot x_f \, dm,$$

where x_f is indicator of the fuzzy set $f \in \mathcal{M}$

Remark 4.2.16 *The notion of an indefinite integral of fuzzy observable x in a fuzzy quantum space was also introduced by Riečan [83], though in a different manner still giving the same value.*

Ergodic Theorems for Fuzzy Quantum Dynamical Systems

In this section, we give analogies of Mesiar's ergodic theorems for the case of fuzzy dynamical system $(\mathbb{X}, \mathcal{M}, m, \tau)$. In the proofs we use the method of F-σ-ideal described above as well as the properties of a σ-homomorphism.

In the following, we generalize the results of Mesiar [56] for a fuzzy quantum space, and we formulate them into the following theorems.

Theorem 4.2.17 *Let $(\mathbb{X}, \mathcal{M}, m, \tau)$ be a given fuzzy dynamical system and $\{x_n\}_{n=1}^{\infty}$ be a sequence of bounded fuzzy observables on fuzzy quantum space $(\mathbb{X}, \mathcal{M})$. Let $K > 0$ be a real constant such that $\|x_n\| \leq K$ for $n = 1,2,\dots$. Suppose $x_n \to o$ almost everywhere in m. Then*

$$\frac{1}{n}\sum_{i=1}^{n} \tau^i \circ x_i \to o \text{ a. e. } [m]. \tag{4.5}$$

Proof. The Boolean sub-σ-algebra $\mathcal{A}$ in Lemma 4.2.7 has a countable generator, hence due to Varadarajan [103], there exists an observable $z \colon \mathcal{B}(\mathbb{R}^1) \to \mathcal{M}/\sim_m$ such that

$$\{z(E) \colon E \in \mathcal{B}(\mathbb{R}^1)\} = \mathcal{A},$$

and a sequence of real-valued Borel functions $\{u_n\}_{n=1}^{\infty}$, such that

$$\overline{x}_n(E) = z(u_n^{-1}(E)), E \in \mathcal{B}(\mathbb{R}^1), n = 1,2,\dots.$$

The sequence $\{u_n\}_{n=1}^{\infty}$ is essentially unique in the following sense: if

$$z(u_n^{-1}(E)) = z(v_n^{-1}(E)), E \in \mathcal{B}(\mathbb{R}^1),$$

then

$$z(\{t: u_n(t) \neq v_n(t)\}) = \overline{0}_{\mathbb{X}}.$$

From the construction of z it follows that $\overline{\tau}$ is z-measurable, *i.e.*

$$\overline{\tau}\Big(z\big(\mathcal{B}(\mathbb{R}^1)\big)\Big) \subset z\big(\mathcal{B}(\mathbb{R}^1)\big).$$

Due to Dvurečenskij and Riečan ([21]) it is possible iff there is a Borel measurable transformation $T: \mathbb{R}^1 \to \mathbb{R}^1$ such that

$$\overline{\tau}(z(E)) = z(T^{-1}(E)), \text{for every } E \in \mathcal{B}(\mathbb{R}^1).$$

Therefore we have

$$\overline{\tau}^k z(E)) = z(T^{-k}(E)), \text{for every } E \in \mathcal{B}(\mathbb{R}^1), k = 1,2,\ldots,$$

and consequently, for every $E \in \mathcal{B}(\mathbb{R}^1)$,

$$\big(\overline{\tau}^k \circ \overline{x}_n\big)(E) = \overline{\tau}^k(\overline{x}_n(E)) = \overline{\tau}^k\big(z(u_n^{-1}(E))\big) = z\big(T^{-1}(u_n^{-1}(E))\big), k = 1,2,\ldots.$$

Moreover,

$$\sigma(x_n) \supset \sigma(\overline{x}_n) \supset \sigma(u_n),$$

and therefore

$$|u_n| \leq \|\overline{x}_n\| \leq \|x_n\| \leq K.$$

Take into account the system $(\mathbb{R}^1, \mathcal{B}(\mathbb{R}^1), \mu_z, T)$, where μ_z is the mapping defined by $\mu_z(E) = \mu(z(E))$, for every $E \in \mathcal{B}(\mathbb{R}^1)$. Then μ_z is a probability measure on $\mathcal{B}(\mathbb{R}^1)$. Moreover, for every $E \in \mathcal{B}(\mathbb{R}^1)$,

$$\mu_z(T^{-1}(E)) = \mu\big(z(T^{-1}(E))\big) = \mu(\overline{\tau}(z(E))) = \mu(z(E)) = \mu_z(E),$$

i.e. the transformation T is μ_z-invariant. This means that the system $(\mathbb{R}^1, \mathcal{B}(\mathbb{R}^1), \mu_z, T)$ is a dynamical system in the classical sense. For observables in a Boolean σ-algebra, there is a well-known way of definition of their sum [103], and the convergence almost everywhere of observables in $\mathcal{M}/\sim_m$ is the same as for fuzzy observables.

By the assumption $x_n \to o$ almost everywhere in m, i.e., for every $\varepsilon > 0$,

$$m\left(\bigvee_{k=1}^{\infty}\bigwedge_{n=k}^{\infty} x_n([-\varepsilon,\varepsilon])\right) = 1.$$

But

$$m\left(\bigvee_{k=1}^{\infty}\bigwedge_{n=k}^{\infty} x_n([-\varepsilon,\varepsilon])\right) = 1, \text{ iff}$$

$$\mu\left(\bigvee_{k=1}^{\infty}\bigwedge_{n=k}^{\infty} \overline{x}_n([-\varepsilon,\varepsilon])\right) = 1, \text{ iff}$$

$$\mu\left(\bigvee_{k=1}^{\infty}\bigwedge_{n=k}^{\infty} z\left(u_n^{-1}([-\varepsilon,\varepsilon])\right)\right) = 1, \text{ iff}$$

$$\mu_z\left(\bigcup_{k=1}^{\infty} \bigcap_{n=k}^{\infty} u_n^{-1}([-\varepsilon,\varepsilon])\right) = 1,$$

which applies iff $u_n \to 0$ a. e. $[\mu_z]$. Then according to Mesiar [56], there holds

$$\frac{1}{n}\Sigma_{i=1}^{n} u_i \circ T^i \to 0 \text{ a. e. } [\mu_z].$$

On the other hand,

$$\frac{1}{n}\Sigma_{i=1}^{n} \tau^i \circ x_i \to o \text{ a. e. } [m]$$

iff

$$\frac{1}{n}\Sigma_{i=1}^{n} \overline{\tau}^i \circ \overline{x}_i \to \overline{o} \text{ a. e. } [\mu],$$

where $\overline{o}(E) = \overline{0}_{\mathbb{X}}$ if $0 \notin E$ and $\overline{o}(E) = \overline{1}_{\mathbb{X}}$ in other cases. The latest convergence is true if and only if

$$\frac{1}{n}\Sigma_{i=1}^{n} \left(u_i \circ T^i\right)(z) \to \overline{o} \text{ a. e. } [\mu],$$

which is possible if and only if

$$\frac{1}{n}\Sigma_{i=1}^{n} u_i \circ T^i \to 0 \text{ a. e. } [\mu_z].$$

Therefore (4.5) is proved.

Theorem 4.2.18 Let $(\mathbb{X}, \mathcal{M}, m, \tau)$ be any fuzzy dynamical system and $\{x_n\}_{n=1}^{\infty}$ be a sequence of fuzzy observables on fuzzy quantum space $(\mathbb{X}, \mathcal{M})$. Let y be a fuzzy observable on $(\mathbb{X}, \mathcal{M})$ such that $o \leq x_n \leq y$ for $n = 1,2,\ldots$. Suppose $x_n \to o$ almost every where in m. Then

$$\frac{1}{n}\sum_{i=1}^{n} \tau^i \circ x_i \to o \text{ a. e. } [m].$$

Proof. We will use the similar arguments as above. Let $\mathcal{A}$ be the minimal Boolean sub-σ-algebra of $\mathcal{M}/\sim_m$ containing all rangers of $\overline{\tau}^k \circ \overline{x}_n$ and $\overline{\tau}^k \circ \overline{y}_n$ for $k = 1,2,\ldots$, $n = 1,2,\ldots$. Then $\overline{\tau}\mathcal{A} \subset \mathcal{A}$ and $\mathcal{A}$ has a countable generator. In view of Varadarajan [103], there is an observable $z: \mathcal{B}(\mathbb{R}^1) \to \mathcal{M}/\sim_m$ such that

$$\{z(E): E \in \mathcal{B}(\mathbb{R}^1)\} = \mathcal{A}.$$

Moreover, there are real-valued Borel functions $u, u_n, n = 1,2,\ldots$ such that

$$\overline{x}_n(E) = z(u_n^{-1}(E)), n = 1,2,\ldots,$$

and

$$\overline{y}(E) = z(u^{-1}(E)),$$

for any $E \in \mathcal{B}(\mathbb{R}^1)$. Denote $v_n = \max(0, u_n)$, $n = 1,2,\ldots$, and $v = \max(0, u_n)$. Then for every $E \in \mathcal{B}(\mathbb{R}^1)$ we have

$$\overline{x}_n(E) = z(v_n^{-1}(E)), n = 1,2,\ldots,$$

and

$$\overline{y}(E) = z(v^{-1}(E)).$$

Let $w_n = \min(0, u_n)$ $n = 1,2,\ldots$, and $w = u$. Then

$$\overline{x}_n(E) = z(w_n^{-1}(E)), n = 1,2,\ldots,$$

and

$$\overline{y}(E) = z(w^{-1}(E)),$$

for any $E \in \mathcal{B}(\mathbb{R}^1)$. Moreover,

$$0_{\mathbb{X}} \leq \|\overline{x}_n\| \leq \|\overline{y}\| \text{ and } 0 \leq w_n \leq w.$$

Similarly as in the proof of Theorem 4.2.17 we take into the dynamical system $(\mathbb{R}^1, \mathcal{B}(\mathbb{R}^1), \mu_z, T)$, where $\mu_z \colon E \to \mu(z(E))$, for every $E \in \mathcal{B}(\mathbb{R}^1)$ is a probability measure on $\mathcal{B}(\mathbb{R}^1)$ such that for every $E \in \mathcal{B}(\mathbb{R}^1)$

$$\mu_z(T^{-1}(E)) = \mu\big(z(T^{-1}(E))\big) = \mu(\overline{\tau}(z(E))) = \mu(z(E)) = \mu_z(E).$$

Thus we have

$$\frac{1}{n}\sum_{i=1}^{n} \tau^i \circ x_i \to o \text{ a. e. } [m] \text{ iff}$$

$$\frac{1}{n}\sum_{i=1}^{n} \overline{\tau}^i \circ \overline{x}_i \to \overline{o} \text{ a. e. } [\mu] \text{ iff}$$

$$\frac{1}{n}\sum_{i=1}^{n} w_i \circ T^i \to 0_{\mathbb{X}} \text{ a. e. } [\mu_z].$$

But the last assertion follows from Theorem 2, which was proved Mesiar in [56].

Remark 4.2.19 *Let x be a fuzzy observable on a fuzzy quantum space $(\mathbb{X}, \mathcal{M})$ and $u(t) = |t|$, $t \in \mathbb{R}^1$, the we put*

$$|x| = u \circ x$$

Corollary 4.2.20 *Let $(\mathbb{X}, \mathcal{M}, m, \tau)$ be any fuzzy dynamical system and $\{x_n\}_{n=1}^{\infty}$ be a sequence of fuzzy observables on fuzzy quantum space $(\mathbb{X}, \mathcal{M})$. Let y be a fuzzy observable on $(\mathbb{X}, \mathcal{M})$ such that $|x_n| \leq y$ for $n = 1,2,\ldots$ If $x_n \to o$ almost everywhere in m, then*

$$\frac{1}{n}\sum_{i=1}^{n} \tau^i \circ x_i \to o \text{ a. e. } [m].$$

Proof. We have $x_n = x_n^+ - x_n^-$, where $x_n^+ = u^+ \circ x_n$, $x_n^- = u^- \circ x_n$, $u^+(t) = \max(0, t)$ and $u^-(t) = -\min(0, t)$, $t \in \mathbb{R}^1$. Then

$$|x| = x^+ + x^-.$$

Applying Theorem 4.2.18 to both sequences $\{x_n^+\}_{n=1}^{\infty}$ and $\{x_n^-\}_{n=1}^{\infty}$ we get what was claimed.

Theorem 4.2.21 (Poincaré recurrence theorem) Let $(\mathbb{X}, \mathcal{M}, m, \tau)$ be a fuzzy dynamical system. Then for every $f \in \mathcal{M}$ we have

$$m\left(f - \bigvee_{j=1}^{\infty} \tau^j(f)\right) = 0.$$

Proof. Put $g = f - \bigvee\limits_{j=1}^{\infty} \tau^i(f)$ Then $\left\{\tau^j(g)\right\}_{j=1}^{\infty}$ are orthogonal fuzzy subsets of $\mathcal{M}$ and therefore by the property (ii) of Definition 3.1.10 we have

$$m\left(\bigvee\limits_{j=1}^{\infty} \tau^j(g)\right) = \Sigma_{j=1}^{\infty} \, m\left(\tau^j(g)\right) = \Sigma_{j=1}^{\infty} \, m(g).$$

Since

$$m\left(\bigvee\limits_{j=1}^{\infty} \tau^j(g)\right) \leq 1 \text{ and } \Sigma_{j=1}^{\infty} \, m(g) \leq 1$$

iff

$$m(g) = 0,$$

the proof is finished.

Definition 4.2.22 *For a sequence $\{f_n\}_{n=1}^{\infty} \subset \mathcal{M}$ we define* $\limsup f_n$ *as follows:*

$$\limsup f_n = \bigwedge\limits_{n=1}^{\infty} \bigvee\limits_{j=n}^{\infty} f_j.$$

Theorem 4.2.23 (Strong Poincaré recurrence theorem) *Let* $(\mathbb{X}, \mathcal{M}, m, \tau)$ *be a fuzzy dynamical system. Then for every $f \in \mathcal{M}$ we have*

$$m\big(f - \limsup \tau^n(f)\big) = 0.$$

Proof. Put $g = \limsup \tau^n(f)$. Then

$$g = f \wedge \left(\bigvee\limits_{n=1}^{\infty}\left(\bigvee\limits_{j=n}^{\infty} \tau^j(f)\right)'\right) =$$

$$= \bigvee\limits_{n=1}^{\infty}\left(f \wedge \left(\bigvee\limits_{j=n}^{\infty} \tau^j(f)\right)'\right) =$$

$$= \bigvee\limits_{n=1}^{\infty}\left(f - \bigvee\limits_{j=n}^{\infty} \tau^j(f)\right) = \bigvee\limits_{n=1}^{\infty} g_n,$$

where

$$g_n = f - \bigvee\limits_{j=n}^{\infty} \tau^j(f), n = 1,2....$$

Applying Theorem 4.2.21 to map $\Pi = \tau^n$ we get for $g_n^* = f - \overset{\infty}{\underset{j=n}{\vee}} \Pi^j(f)$

$$m(g_n^*) = 0.$$

But $g_n \leq g_n^*$, and therefore by property (v) of Theorem 3.3.1 we get

$$m(g_n) = 0, n = 1,2\ldots..$$

Since any fuzzy P-measure is a σ-subadditive function, we obtain

$$m(g) = m\left(\overset{\infty}{\underset{n=1}{\vee}} g_n\right) \leq \Sigma_{n=1}^{\infty} m(g_n) = 0.$$

4.3 THE HAHN-JORDAN DECOMPOSITION OF A FUZZY QUANTUM SPACE

In this section we show the existence of Hahn decomposition of a fuzzy quantum space and we also prove the existence of Jordan and Lebesgue decomposition of a signed measure on fuzzy quantum space $(\mathbb{X}, \mathcal{M})$. The Lebegue decomposition theorem for the fuzzy case was proved in [51] and [101].

In the beginning of this section we recall basic concepts which are necessary for defining decompositions on fuzzy quantum space $(\mathbb{X}, \mathcal{M})$.

Definition 4.3.1 *A mapping* $m \colon \mathcal{M} \to \mathbb{R}^1$ *such that if* $\{f_k\}_{k=1}^{\infty}$ *is a sequence of pairwise orthogonal fuzzy subsets from* $\mathcal{M}$, *i.e.* $f_i \perp f_j$, $(f_i \leq 1_{\mathbb{X}} - f_j)$, *whenever* $i \neq j$, *then*

$$m\left(\overset{\infty}{\underset{i=1}{\vee}} f_i\right) = \Sigma_{i=1}^{\infty} m(f_i),$$

m is called non-negative signed measure on fuzzy quantum space $(\mathbb{X}, \mathcal{M})$.

In particular, measure m with property $m(1_{\mathbb{X}}) = 1$ is a so-called fuzzy probability measure, in Piasecki's terminology m is a fuzzy P-measure [70].

Theorem 4.3.2 *Let m be a signed measure of* $(\mathbb{X}, \mathcal{M})$. *Then every system of mutually orthogonal fuzzy subsets* $f \in \mathcal{M}$ *such that* $m(f) > 0$ *($m(f) < 0$) is countable.*

Proof. *Denote* $\mathcal{G} \subset \mathcal{M}$ *be a system of mutually orthogonal sets* $f \in \mathcal{M}$ *for which* $m(f) > 0$. *Let us put*

$$\mathcal{G}_r = \left\{ f \in \mathcal{G}; \ m(f) > \frac{1}{r} \right\}, r = 1, 2, \ldots.$$

Evidently,

$$\mathcal{G} = \bigcup_{r=1}^{\infty} \mathcal{G}_r. \tag{4.6}$$

It is clear, that the system $\mathcal{G}_r$ is finite for every r. Namely, in the opposite case there exists a sequence $\{f_k\}_{k=1}^{\infty}$ of mutually orthogonal fuzzy sets from $\mathcal{G}_r$, *i.e.* such that $m(f_k) > \frac{1}{r}$ for $k = 1, 2, \ldots$. Calculate

$$m\left(\bigvee_{k=1}^{\infty} f_k \right) = \sum_{k=1}^{\infty} (f_k) \geq \sum_{k=1}^{\infty} \frac{1}{r} = \infty,$$

which gives a contradiction. Therefore, from the property (4.6) we get that $\mathcal{G}$ is countable.

Definition 4.3.3 Let m be any signed measure defined on a soft fuzzy σ-algebra $\mathcal{M}$ of fuzzy subsets of $\mathbb{X}$. A fuzzy set $f \in \mathcal{M}$ is called positive (negative) with respect to a signed measure m if

$$m(f \wedge g) \geq 0 \ (m(f \wedge g) \leq 0)$$

for every fuzzy set $g \in \mathcal{M}$.

It is evident that a fuzzy set $g \in \mathcal{M}$ is positive (negative) with respect to a signed measure m if and only if for every $f \in \mathcal{M}$ such that $f \leq g$ it holds that $m(f) \geq 0$ $(m(f) \leq 0)$.

Definition 4.3.4 (Hahn decomposition) An ordered couple (f, g) where f is positive and g is negative set with respect to m such that $g = f'$, is called Hahn decomposition of $(\mathbb{X}, \mathcal{M})$ with respect to a signed measure m.

Theorem 4.3.5 (Hahn decomposition) Let m be a signed measure on $\mathcal{M}$, then the Hahn decomposition of a fuzzy quantum space $(\mathbb{X}, \mathcal{M})$ with respect to m exists.

Proof. Without loss of generality we can assume that there is a maximal system $\mathcal{G}$ of mutually orthogonal sets $g \in \mathcal{M}$, which are negative $(m(g) < 0)$ with respect to m (In opposite case m is a measure and we put $f = 1_{\mathbb{X}}, g = 0_{\mathbb{X}}$). By Theorem

4.3.2 the system $\mathcal{G}$ is countable. Let $g = \vee \ \{f \colon f \in \mathcal{G}\} \in \mathcal{M}$ and let $f = g' \in \mathcal{M}$, then

$$m(g) = m(\vee f) = \Sigma \ m(f) < 0.$$

For every $e \in \mathcal{M}$, one holds

$$m(e \wedge g) = m\left(e \wedge \underset{i}{\vee} f_i\right) = m\left(\underset{i}{\vee} (e \wedge f_i)\right) = \Sigma_i \ m(e \wedge f_i) \le 0,$$

that is, g is negative.

Now we show that f is positive. Let f be not positive. Then there is $e_0 \in \mathcal{M}$, such that $e_0 \le f$, $m(e_0) < 0$. We denote by $\mathcal{G}_0$ the maximal system of mutually orthogonal sets $l \in \mathcal{M}$, $l \le e_0$, $m(l) > 0$. In view Theorem 4.3.2, the system $\mathcal{G}_0$ is countable.

Let $l_0 = \vee \ \{l \colon l \in \mathcal{G}_0\}$, then $m(l_0) > 0$, $l_0 \le e_0$. We show, that $l_0 \wedge e_0'$ is negative, because it does contain any set of a positive measure. From the equality (iii) of Theorem 3.3.1,

$$m(e_0) = m(e_0 \wedge l_0') + m(l_0),$$

which entails

$$m(e_0 \wedge l_0') < 0.$$

Then the set $e_0 \wedge l_0'$ is negative and $(e_0 \wedge l_0') \perp g$, which is a contradiction with the maximality of a system $\mathcal{G}_0$.

Lemma 4.3.6 Let (f_1, g_1), (f_2, g_2) be two Hahn decomposition of $(\mathbb{X}, \mathcal{M})$ with respect to m. Then

$$m(s \wedge f_1) = m(s \wedge f_2),$$

$$m(s \wedge g_1) = m(s \wedge g_2),$$

for any $s \in \mathcal{M}$.

Proof. Since

$$s \wedge f_1 \wedge f_2' \le f_1,$$

we have

$$m(s \wedge f_1 \wedge f_2') \ge 0.$$

The inequality

$$s \wedge f_1 \wedge f_2' \leq g_2,$$

implies

$$m(s \wedge f_1 \wedge f_2') \leq 0.$$

Hence

$$m(s \wedge f_1 \wedge f_2') = 0.$$

Analogously, we obtain

$$m(s \wedge f_2 \wedge f_1') = 0.$$

Therefore by property (vii) of Theorem 3.3.1 and additivity of m we get

$$m(s \wedge f_1) = m(s \wedge f_1 \wedge (f_2 \wedge f_2')) = m(s \wedge f_1 \wedge f_2) + m(s \wedge f_1 \wedge f_2') =$$

$$= m(s \wedge f_1 \wedge f_2),$$

$$m(s \wedge f_2) = m(s \wedge f_2 \wedge (f_1 \wedge f_1')) = m(s \wedge f_2 \wedge f_1) + m(s \wedge f_2 \wedge f_1') =$$

$$= m(s \wedge f_1 \wedge f_2),$$

i.e.

$$m(s \wedge f_1) = m(s \wedge f_2)$$

The equality

$$m(s \wedge g_1) = m(s \wedge g_2)$$

may be proved the same way.

Let m be a signed measure on $\mathcal{M}$ and (f, g) be any Hahn decomposition of $(\mathbb{X}, \mathcal{M})$ with respect to m. Let us define the mapping m^+ and m^- by the equalities

$$m^+(e) = m(e \wedge f),$$

$$m^-(e) = -m(e \wedge g),$$

for every $e \in \mathcal{M}$. It easy to verify that the mapping m^+ and m^- are measures on $(\mathbb{X}, \mathcal{M})$. The measures m^+ and m^- are independent of Hahn decomposition which follows immediately from the Lemma 4.3.6. Moreover, for any $e \in \mathcal{M}$, it holds

$$m(e) = m^+(e) - m^-(e). \tag{4.7}$$

Definition 4.3.7 (Jordan decomposition) *The formula (4.7) is called a Jordan decomposition of a signed measure m. The measure m^+ (m^-) is said to be positive (negative) part of m. The measure $|m|$ defined for any $e \in \mathcal{M}$ by*

$$|m|(e) = m^+(e) + m^-(e) \tag{4.8}$$

is called the total variation of a signed measure m.

We will prove the next theorem in two ways:

1.　using the method of representation of fuzzy observables, which we introduced in this chapter,

2.　using method of F-σ-ideals, which we described in the third chapter.

Theorem 4.3.8 *Let m be a fuzzy measure and let x be a fuzzy observable on a fuzzy quantum space $(\mathbb{X}, \mathcal{M})$ such that*

$$m(|x|): \int_{\mathbb{R}^1} |t| dm_x(t) < \infty.$$

Then the mapping ν defined *via*

$$\nu(f) = \int_f x \, dm, f \in \mathcal{M},$$

is a signed measure on $(\mathbb{X}, \mathcal{M})$, where positive and negative parts, ν^+ and ν^-, are defined as follows

$$\nu^+(f) = \int_f x^+ dm, f \in \mathcal{M},$$

$$\nu^-(f) = -\int_f x^- dm, f \in \mathcal{M},$$

where $x^+ = h^+ \circ x$, $x^- = h^- \circ x$ and $h^+(t) = \max(t, 0)$, $h^-(t) = -\min(t, 0)$, $t \in \mathbb{R}^1$.

Proof. 1. Let $x^+ = h^+ \circ x$, $x^- = h^- \circ x$. Then (f, g) is a Hahn decomposition of $(\mathbb{X}, \mathcal{M})$ defined *via*

$$f = x([0, \infty)),$$

$$g = x((-\infty, 0)).$$

Let $x^+(\mathbb{R}^1) = x((h^+)^{-1}(\mathbb{R}^1)) = x(\mathbb{R}^1)$. We can show that

$$m(e \wedge f) \geq 0, m(e \wedge g) \leq 0 \text{ for any } e \in \mathcal{M}.$$

We denote

$$1_k = (f \vee f') \wedge (e \vee e') \wedge x(\mathbb{R}^1),$$

where k is an integer or ∞ and we define new observables of $(\mathbb{X}, \mathcal{M})$:

$$\overline{x}(E) = x(E) \wedge 1_k \vee 0_k$$

$$\overline{x}_f(E) = x_{\overline{f}}(E)$$

$$\overline{x}_e(E) = x_{\overline{e}}(E)$$

$$\overline{x}_{f \wedge e}(E) = x_{\overline{f \wedge e}}(E) \wedge 1_k \vee 0_k, E \in \mathcal{B}(\mathbb{R}^1),$$

where

$$\overline{f} = f \wedge 1_k \vee 0_k,$$

$$\overline{e} = e \wedge 1_k \vee 0_k.$$

Then

$$\overline{x}_{f \wedge e} = x_{\overline{f \wedge e}} = x_{\overline{f}} \cdot x_{\overline{e}}$$

and due to Dvurečenskij [10], there exists a mapping $\phi: \mathbb{X}_{1_k} \to \mathbb{R}^1$ such that $\mathbb{X}_{1_k}(\overline{x}(E)) = \phi^{-1}(E)$ for every $E \in \mathcal{B}(\mathbb{R}^1)$, where

$$\mathbb{X}_{1_k}(g) = \{s \in \mathbb{X}_{1_k} : g(s) = 1_k(s)\},$$

$$\mathbb{X}_{1_k} = \left\{s \in \mathbb{X} : 1_k(s) \neq \frac{1}{2}\right\}$$

and g is a fuzzy set from $\mathcal{M}$ for which:

$$(g \vee g')(s) = 1_k(s) \text{ for any } s \in \mathbb{X}.$$

Due to Dvurečenskij [10], a σ-algebra of crisp subsets $\mathbb{X}_{1_k}(1_{\mathbb{X}})$ is generated by the system $\mathcal{A}_{1_k}$ of all fuzzy sets $\mathbb{X}_{1_k}$ and mapping $\mu_{1_k}: \mathcal{A}_{1_k} \to [0,1]$ is defined *via* $\mu_{1_k}(\mathbb{X}_{1_k}(g)) = m(g)$, $g \in \mathcal{A}_{1_k}$ is a probability measure on $\mathcal{A}_{1_k}$. It may be proved that

$$\mathbb{X}_{1_k}(\overline{x}_f(E)) = I_A^{-1}(E)$$

$$\mathbb{X}_{1_k}(\overline{x}_e(E)) = I_F^{-1}(E)$$

for any $E \in \mathcal{B}(\mathbb{R}^1)$, where $A = \{s \in \mathbb{X}_{1_k}: \phi(s) \geq 0\}$ and F is (unique) crisp subsets of $\mathcal{A}_{1_k}$. Calculate

$$v^+(e) = v(e \wedge f) = \int_{e \wedge f} x\, dm = \int_{A \cap F} \phi(s)\, d\mu_{1_k}(s) \geq 0.$$

Analogously we have

$$v^-(e) = -v(e \wedge f) = -\int_{e \wedge f} x\, dm = -\int_{A \cap F} \phi(s)\, d\mu_{1_k}(s) \leq 0.$$

In same way we prove

$$v^-(e) = -\int_e x^-\, dm,$$

which finishes the first proof.

2. In the Definition 3.3.4 we defined the equivalence relation $f \sim_m g$ as follows: for every $f, g \in \mathcal{M}$, $f \sim_m g$ iff $m(f \wedge g') = m(g \wedge f') = 0$. The system

$$\overline{\mathcal{M}} = \mathcal{M}/\sim_m = \{\overline{f}: f \in \mathcal{M}\},$$

where

$$\overline{f} = \{g \in \mathcal{M}: g \sim_m f\},$$

is a Boolean σ-algebra. By Theorem 3.3.6 the mapping μ from $\overline{\mathcal{M}}$ into $\mathbb{R}^1$ defined as

$$\mu(\overline{f}) = m(f),$$

for every $\overline{f} \in \overline{\mathcal{M}}$, is a signed measure on the Boolean σ-algebra $\overline{\mathcal{M}}$. We define a σ-homomorphism $\overline{x}$ from $\mathcal{B}(\mathbb{R}^1)$ into $\overline{\mathcal{M}}$ as follows: $\overline{x}: E \to \overline{x}(E)$, $E \in \mathcal{B}(\mathbb{R}^1)$.

From the Loomis-Sikorski theorem [97] applied to the Boolean σ-algebra $\overline{\mathcal{M}}$ it follows that there exists a measurable space $(\Omega, \mathcal{S})$ ($\mathcal{S}$ is σ-algebra of subsets of Ω) and a σ-homomorphism h from $\mathcal{S}$ onto $\overline{\mathcal{M}}$. In view of Varadarajan [103], there is an $\mathcal{S}$-measurable, real-valued function ψ such that

$$\overline{x}(E) = h(\psi^{-1}(E)), E \in \mathcal{B}(\mathbb{R}^1).$$

Let us define the mapping $P_m \colon \mathcal{S} \to [0,1]$ in the following way:

$$P_m(A) = \mu(h(A)), A \in \mathcal{S}.$$

It is simple to verify, that P_m is a probability measure on $\mathcal{S}$. Hence, for any $e \in \mathcal{M}$, there is $G \in \mathcal{S}$ such that $\overline{e} \in h(G)$. Moreover, if we put

$$A^+ = \{\omega \in \Omega \colon \psi(\omega) \geq 0\},$$

$$A^- = \{\omega \in \Omega \colon \psi(\omega) < 0\},$$

then

$$\overline{f} = h(A^+) \text{ and } \overline{g}' = h(A^-).$$

Calculate

$$\nu^+(f) = \nu(e \wedge f) = \int_{e \wedge f} x dm = \int_{\overline{e} \wedge \overline{f}} \overline{x} d\mu = \int_{G \cap A^+} \psi(\omega) dP_m(\omega) \geq 0.$$

$$\nu^-(f) = -\nu(e \wedge f) = -\int_{e \wedge f} x dm = -\int_{G \cap A^-} \psi(\omega) dP_m(\omega) \leq 0.$$

4.4 THE LEBESGUE DECOMPOSITION THEOREM FOR A FUZZY QUANTUM SPACE

In this section, we present the fuzzy analogy of the Lebesgue decomposition theorem. For this we shall need some auxiliary assertions.

The following Lemma is proved in [62].

Lemma 4.4.1 (Zorn) *Let S be a nonempty partially ordered set. Let for every ordered subset $L \subset S$ there exists $s \in S$ such that $x \leq s$ for all $x \in L$. Then there exists a maximal element $d \in S$ (i.e. such element $d \in S$ that $\{x \in S; d < x\} = \emptyset$.)*

Proposition 4.4.2 *Let $\mathcal{E}$ be any system of fuzzy subsets of $\mathbb{X}$. Then there exists a maximal subsystem $\mathcal{E}_0 \subset \mathcal{E}$ of mutually orthogonal fuzzy sets (i.e. such that $\{f \in \mathcal{E}; f \perp g \; forall \; g \in \mathcal{E}_0\} = \emptyset.)$*

Proof. Denote by $\overline{\mathcal{E}}$ the system consisting of all subsystems of $\mathcal{E}$ elements of which are mutually orthogonal subsets. The system $\overline{\mathcal{E}}$ is partially ordered by the relations of inclusion. Let $\mathcal{L} \subset \overline{\mathcal{E}}$ be any ordered system. Then there exists an element $\mathcal{E}_{\mathcal{L}} \in \overline{\mathcal{E}}$ such that $x \subset \mathcal{E}_{\mathcal{L}}$ for all $x \in \mathcal{L}$. It suffices to put

$$\mathcal{E}_{\mathcal{L}} = \bigcup_{x \in \mathcal{L}} x.$$

This means that $\overline{\mathcal{E}}$ satisfies the assumptions of Zorn's lemma. Therefore, there exists a maximal system $\mathcal{E}_0 \in \overline{\mathcal{E}}$.

Definition 4.4.3 *Let $(\mathbb{X}, \mathcal{M})$ be any fuzzy quantum space. Le m be a signed measure of $(\mathbb{X}, \mathcal{M})$ and n be a measure of $(\mathbb{X}, \mathcal{M})$. We say that a signed measure m is dominated by a measure n (and we write $m \ll n$) if $n(f) = 0$ implies $m(f) = 0$ for every $f \in \mathcal{M}$.*

Remark 4.4.4 *Let m be a fuzzy P-measure of $(\mathbb{X}, \mathcal{M})$ and x be a fuzzy observable of $(\mathbb{X}, \mathcal{M})$ such that*

$$\int_{\mathbb{R}^1} |t| dm_x(t) < \infty.$$

We defined (Definition 4.2.15) an integral on a fuzzy observable x of $(\mathbb{X}, \mathcal{M})$ over a fuzzy set $f \in \mathcal{M}$ as follows

$$v(f) = \int_f x dm := \int x \cdot x_f dm, \qquad (4.9)$$

where x_f is the indicator of the fuzzy set f.

Let us consider the signed measure v of $(\mathbb{X}, \mathcal{M})$ defined by (4.9). Dvurečenskij in [10] proved that, for every $f \in \mathcal{M}$, $m(f) = 0$ implies $v(f) = 0$. So, the signed measure v is dominated by the P-measure m.

Theorem 4.4.5 *Let m be a signed measure of $(\mathbb{X}, \mathcal{M})$ and n be a measure of $(\mathbb{X}, \mathcal{M})$. Then the following assertions are equivalent:*

$$m \ll n, \qquad (4.10)$$

$$m^+ \ll n, m^- \ll n, \tag{4.11}$$

$$|m| \ll n. \tag{4.12}$$

Proof. Let (f, g) be a Hahn decomposition of $(\mathbb{X}, \mathcal{M})$ with respect to m. Let $m \ll n$. If $e \in \mathcal{M}$ such that $n(e) = 0$, then the monotonicity of n implies $n(e \wedge f) = 0$ By means of (4.10) we obtain $m(e \wedge f) = 0$, *i.e.* $m^+(e) = 0$, which gives that $m^+ \ll n$. Analogously it may be proved that $m^- \ll n$.

Now, let (4.11) hold. From equality (4.8) it is clear, that (4.12) holds. If $e \in \mathcal{M}$ such that $n(e) = 0$, then $|m|(e) = m^+(e) + m^-(e) = 0$. Since m^+ and m^- are measures, that last equality implies $m^+(e) = m^-(e) = 0$, which proves that $m \ll n$. The proof is finished.

Let m be a signed measure on $(\mathbb{X}, \mathcal{M})$ and n be a measure of $(\mathbb{X}, \mathcal{M})$. We will write $m \ll_\varepsilon n$ if to every $\varepsilon > 0$ there exists $\delta > 0$ such that for every $e \in \mathcal{M}$ the inequality $n(e) < \delta$ implies $|m(e)| < \varepsilon$.

Theorem 4.4.6 *Let m be a signed measure of $(\mathbb{X}, \mathcal{M})$ and n be a measure of $(\mathbb{X}, \mathcal{M})$. Then m is dominated by n if and only if $m \ll_\varepsilon n$.*

Proof. Let $m \ll_\varepsilon n$. If $e \in \mathcal{M}$ such that $n(e) = 0$, then $n(e) < \delta$ for every $\delta > 0$ and hence $|m(e)| < \varepsilon$ for every $\varepsilon > 0$. We get $m(e) = 0$. This means that $m \ll n$.

Let us prove the opposite implication. Let $m \ll n$. We will prove that $m^+ \ll_\varepsilon n$. Let $m \ll n$ and $m^+ \ll_\varepsilon n$ does not hold, *i.e.* there exists $\varepsilon > 0$ such that for $\delta_k = \frac{1}{2^k}$ $(k = 1,2,...)$ there exists $e_k \in \mathcal{M}$ for which $n(e_k) < \frac{1}{2^k}$ and $m^+(e_k) \geq \varepsilon$. Put

$$e = \bigwedge_{k=1}^{\infty} \bigvee_{r=k}^{\infty} e_r.$$

If we denote $l_k = \bigvee_{r=k}^{\infty} l_r$, then $e = \bigwedge_{k=1}^{\infty} l_k$ and $\{l_k\} \searrow e$. By by means of monotonicity of the measure m^+, we get

$$m^+(l_k) = m^+\left(\bigvee_{r-k}^{\infty} e_r\right) \geq m^+(e_k) \geq \varepsilon, k = 1,2,...$$

and hence by Theorem 3.3.1 property (iv) we have

$$m^+(e) = \lim_{k \to \infty} m^+(l_k) \geq \varepsilon.$$

On the other hand,

$$n(e) = n \left(\bigwedge_{k=1}^{\infty} \bigvee_{r=k}^{\infty} e_r \right) \leq n \left(\bigvee_{r-k}^{\infty} e_r \right) \leq \sum_{r=k}^{\infty} n(e_r) \leq \sum_{r=k}^{\infty} \frac{1}{2^k} = \frac{1}{2^{k-1}}$$

for $k = 1, 2, \ldots$, when we used the monotonicity and the σ-subadditivity of n.

Therefore $n(e) = 0$ and $m^+(e) \neq 0$. If we take account of Theorem 4.4.6, we see that this result presents a contradiction. Analogously, we obtain that $m^- \ll_\varepsilon n$. If $m^+ \ll_\varepsilon n$ and $m^- \ll_\varepsilon n$, then it is easy to see that $m \ll_\varepsilon n$.

Definition 4.4.7 *Let m, n be two measures on a quantum space $(\mathbb{X}, \mathcal{M})$. We say that m is singular with respect to n (and write $m \perp n$) if there exists two fuzzy sets $f, g \in \mathcal{M}$, $f = g'$, such that $m(e \wedge f) = n(e \wedge g) = 0$ for every $e \in \mathcal{M}$.*

The following theorem is a fuzzy analogy of the Lebesgue decomposition theorem.

Theorem 4.4.8 (Lebesgue decomposition theorem) *Let m and n be two measures of $(\mathbb{X}, \mathcal{M})$. Then there exists measures m_1, m_2 on $\mathcal{M}$ such that $m_1 \ll n$, $m_2 \perp n$ and*

$$m(e) = m_1(e) + m_2(e)$$

for every $e \in \mathcal{M}$.

Proof. Denote by $\mathcal{E}$ the system of fuzzy sets $e \in \mathcal{M}$ for which $n(e) = 0$ and $m(e) > 0$. It is easy to see that $\mathcal{E}$ is a nonempty set. Namely, in opposite case, m is dominated by n and it suffices to put $m_1 = m$, $m_2 = 0$.

By Proposition 4.4.2, there exists a maximal system $\mathcal{E}_0$ of mutually orthogonal fuzzy sets from $\mathcal{E}$. By means of Theorem 4.3.2, $\mathcal{E}_0$ is countable, *i.e.* $\mathcal{E}_0 = \{e_k\}_{k=1}^{\infty}$. Put $e_0 = \bigvee_{k=1}^{\infty} e_k$. Then $e_0 \in \mathcal{M}$.

Let define the mapping $m_1: \mathcal{M} \to \mathbb{R}^1$ by $m_1(e) = e \wedge e_0$, for every $e \in \mathcal{M}$. It is evident that m_1 is a measure of $(\mathbb{X}, \mathcal{M})$.

Now, let $e \in \mathcal{M}$ be such that $n(e) = 0$. Then the monotonicity of n implies $n(e \wedge e_0') = 0$. Since

$$e \wedge e_{0\prime} \leq e_{0\prime} = \bigwedge_{k=1}^{\infty} e_{k\prime} \leq e_{k\prime}$$

for $k = 1,2,\dots$, *i.e.* $e \wedge e_{0\prime} \perp e_k$ for $k = 1,2,\dots$, we obtain

$$m_1(e) = m(e \wedge e_{0\prime}) = 0.$$

(In the case that $m(e \wedge e_{0\prime}) > 0$, we get the contradiction to the maximality of the system $\mathcal{E}_0$.) This means that m_1 is dominated by n.

Further, let us define the measure $m_2 : \mathcal{M} \to \mathbb{R}^1$ by $m_2(e) = e \wedge e_0)$. We will show that m_2 is singular with respect to n. Put $f = e_0$, $g = e_{0\prime}$. Then by the monotonicity and σ-additivity of n we gwt

$$m(e \wedge f) = n(e \wedge e_0) \leq n\left(\bigvee_{k=1}^{\infty} e_k\right) = \sum_{k=1}^{\infty} n(e_k) = 0$$

for every $e \in \mathcal{M}$. By means of monotonicity of m and Theorem 3.1 we have

$$m_2(e \wedge g) = m(e \wedge e_{0\prime} \wedge e_0) \leq (e_{0\prime} \wedge e_0) = 0$$

for every $e \in \mathcal{M}$. Finally for every for every $e \in \mathcal{M}$,

$$m(e) = m(e \wedge g) = m(e \wedge (e_0 \wedge e_{0\prime})) =$$

$$= m(e \wedge e_0) + m(e \wedge e_{0\prime}) =$$

$$= m_1(e) + m_2(e).$$

CHAPTER 5

Statistical Applications

Abstract: In this chapter we compare two methods applied to reduce the dimensionality of data sets. The first method is Principal component analysis, and second method is Factor analysis. We present these methods on data from Atanassov's intuitionistic fuzzy sets [6]. Earlier we construct an example of applying these methods. The calculations are performed in program R.

Keywords: Principal component analysis, Factor analysis, Atanassov intuitionistic fuzzy sets, Membership function, Non-membership function, Hesitation margin, Correlation, Correlation matrix, Eigenvalues.

5.1 INTRODUCTION

IF sets are mentioned in Chapter 2. We defined membership function and non-membership function. Now we define the hesitation margin of IF set $A = (\mu_A, \nu_A)$ in the following way:

$$\pi_A(\omega) = 1 - \mu_A(\omega) - \nu_A(\omega), \tag{5.1}$$

$\omega \in \Omega$. It is obvious that $0 \leq \pi_A(\omega) \leq 1$ for every $\omega \in \Omega$.

Data from the Atanassov intuitionistic fuzzy set are better characterized by the nature of the studied components. In the classical case we examine a sample of the one-sided point of view, but if sample is from IF sets then this sample is examined from three perspectives (membership function, non-membership function and hesitation margin of IF set). Atanassov intuitionistic fuzzy sets was presented in [98].

5.2 CORRELATION BETWEEN THE ATANASSOV IF-SETS

Correlation between the Atanassov IF-sets (denote A-IFSs) was introduced by Szmidt and Kacprzyk [38] in 2010.

Let A, B be A-IFSs defined on $\Omega = \{\omega_1, \omega_2, \ldots, \omega_n\}$. The sets A, B are characterized by a sequence of pairs:

$[(\mu_A(\omega_1), \nu_A(\omega_1), \pi_A(\omega_1)), (\mu_B(\omega_1), \nu_B(\omega_1), \pi_B(\omega_1))],$

$[(\mu_A(\omega_2), \nu_A(\omega_2), \pi_A(\omega_2)), (\mu_B(\omega_2), \nu_B(\omega_2), \pi_B(\omega_2))],$

...

$[(\mu_A(\omega_n), \nu_A(\omega_n), \pi_A(\omega_n)), (\mu_B(\omega_n), \nu_B(\omega_n), \pi_B(\omega_n))],$

which correspond to the membership values, non-membership values and hesitation margins of A and B.

Definition 5.2.1 (Szmidt, Kacprzyk, Bujnowski [8]) *The correlation coefficient* $r_{A-IFS}(A, B)$ *between two A-IFSs A and B in Ω is:*

$$r_{A-IFS}(A, B) = \frac{1}{3}\left(r_1(A, B) + r_2(A, B) + r_3(A, B)\right) \qquad (5.2)$$

where

$$r_1(A, B) = \frac{\sum_{i=1}^{n}(\mu_A(\omega_i)-\bar{\mu}_A)(\mu_B(\omega_i)-\bar{\mu}_B)}{\left(\sum_{i=1}^{n}(\mu_A(\omega_i)-\bar{\mu}_A)^2\right)^{0.5}\left(\sum_{i=1}^{n}(\mu_B(\omega_i)-\bar{\mu}_B)^2\right)^{0.5}}, \qquad (5.3)$$

$$r_2(A, B) = \frac{\sum_{i=1}^{n}(\nu_A(\omega_i)-\bar{\nu}_A)(\nu_B(\omega_i)-\bar{\nu}_B)}{\left(\sum_{i=1}^{n}(\nu_A(\omega_i)-\bar{\nu}_A)^2\right)^{0.5}\left(\sum_{i=1}^{n}(\nu_B(\omega_i)-\bar{\nu}_B)^2\right)^{0.5}}, \qquad (5.4)$$

$$r_3(A, B) = \frac{\sum_{i=1}^{n}(\pi_A(\omega_i)-\bar{\pi}_A)(\pi_B(\omega_i)-\bar{\pi}_B)}{\left(\sum_{i=1}^{n}(\pi_A(\omega_i)-\bar{\pi}_A)^2\right)^{0.5}\left(\sum_{i=1}^{n}(\pi_B(\omega_i)-\bar{\pi}_B)^2\right)^{0.5}}. \qquad (5.5)$$

where

$$\bar{\mu}_A = \frac{1}{n}\sum_{i=1}^{n}\mu_A(\omega_i), \quad \bar{\nu}_A = \frac{1}{n}\sum_{i=1}^{n}\nu_A(\omega_i), \quad \bar{\pi}_A = \frac{1}{n}\sum_{i=1}^{n}\pi_A(\omega_i),$$

$$\bar{\mu}_B = \frac{1}{n}\sum_{i=1}^{n}\mu_B(\omega_i), \quad \bar{\nu}_B = \frac{1}{n}\sum_{i=1}^{n}\nu_B(\omega_i), \quad \bar{\pi}_B = \frac{1}{n}\sum_{i=1}^{n}\pi_B(\omega_i)$$

The correlation coefficient (5.2) depends on two factors:

- the amount of information expressed by the membership and non-membership degrees (5.3), (5.4)

- the reliability of information expressed by the hesitation margins (5.5).

The correlation coefficient (5.2) possesses the following properties [8]:

1. $r_{A-IFS}(A, B) = r_{A-IFS}(B, A)$,

2. If $A = B$ then $r_{A-IFS}(A, B) = 1$,

3. $|r_{A-IFS}(A, B)| \leq 1$.

This properties are satisfied by its every component (5.3) - (5.5).

5.3 PRINCIPAL COMPONENT ANALYSIS AND FACTOR ANALYSIS FOR A-IFS DATA

Principal Component Analysis

Principal component analysis (PCA) was introduced in 1901 by Karl Pearson. The aim of the method is to transform input multidimensional data in order to make output data the most important linear directions, while the least significant directions are ignored. We extract characteristic directions of the original data, and at the same time the dimension of data is reduced. This method is one of the basic methods of data compression - original n-tuple of variables can represent a smaller number m of variables. The new number of variables explains sufficiently large part of variability of the original data set. The system of new variables (called principal components) consists of linear combinations of the original variables. The first principal component describes the largest part of variability of the original data set, other principal components contribute to the overall variance a smaller share. All the pairs of principal components are orthogonal to each other.

Basic steps of PCA:

- construct of correlation matrix of the source data,

- calculation of eigenvalues of the correlation matrix and eigenvalue in descending order ($\lambda_1 > \ldots > \lambda_n$),

- calculation of eigenvectors of the correlation matrix corresponding to eigenvalues ($v_1, \ldots, v_n$),

- calculation of the variability of the source data (σ^2),

- determine the number of principal components based on variability, which are sufficient to represent the original variables,

- convert the source data into new basis.

Determining the number of principal components:

- At its own discretion about the need to preserve information (eigenvalues which explain eg. 90% variability).

- Kaiser rule: use the principal component whose eigenvalue is greater than 1.

- Use the principal component, which together explains at least 70% of the total variance.

- Scree Plot - graph of eigenvalues. It is recommended to use a number of factors that are before elbow (break point) on the graph.

Factor Analysis

Factor analysis was introduced in 1904 by Charles Edward Spearman. This method allows one to define new variables based on a set of original variables. It enables one to identify covert (latent) causes which are the source of data variability. Using the latent variables enables one to reduce the number of variables while maintaining maximal amount of information, and to find the relationship between observable causes and new variables (factors). Assuming that input variables are mutually correlated, the same amount of information can be described by a smaller number of variables. In the final solution each original variable should correlate with as small number of factors as possible, and the number of factors should be minimal.

Factor loadings reflect the effect of the k-th common factor for j-th random variable.

For estimation of the factor loadings several methods are used and they are referred to methods of extraction factors. We use Principal Component Analysis.

Determining of the number of common factors:

- The criterion of the eigenvalues - the factors, which have their eigenvalues $\lambda > 1$ are significant. If the number of variables from 20 to 50, then the rule is reliable.

- Variance explained criteria - common factors should explain the most of the total variability.

- Scree plot.

Basic steps of FA:

- selection of data,

- determining the number of common factors,

- estimation of the parameters,

- rotation of factors (Varimax Method = orthogonal rotation),

- estimate of elements of the factor matrix (= matrix of factor loadings)

The result of the Factor analysis is matrix of factor loadings. If factor loading is high (> 0.5), than this factor is statistically significant.

Factor analysis is related to principal component analysis, but the two are not identical. Latent variable models, including factor analysis, use regression modelling techniques to test hypotheses producing error terms, while PCA is a descriptive statistical technique [7].

The main idea of both methods: FA and PCA are trying to reduce the dimensionality of the data group.

Example

Job position

We have 20 candidates to job. In the selection of candidates four criteria were rated:

A - Qualifying

B - Communication

C - Independent

D - Skill

Each criterion was evaluated 2 times. How many percent is the criterion met for each participant and how many percent is the criterion not met. The results are in following Table **5.1**.

Table 5.1. Designation: m = meets, nm = not meets.

	A (%)		B (%)		C (%)		D (%)	
	m	nm	m	nm	m	nm	m	nm
1	32	50	64	21	67	15	70	10
2	61	20	37	55	65	20	65	25
3	59	25	40	50	43	22	40	30
4	36	50	62	32	35	40	40	40
5	62	20	46	50	40	38	20	75
6	52	35	84	10	87	5	80	5
7	76	15	52	35	80	12	75	10
8	89	5	70	15	73	10	70	15
9	59	37	85	5	58	21	64	22
10	53	47	40	54	52	37	52	37
11	78	12	62	37	60	18	62	28
12	90	5	54	38	78	5	82	10
13	65	24	30	51	82	2	75	13
14	58	40	15	65	62	15	59	24
15	75	12	65	17	40	45	80	12
16	32	60	85	8	54	38	30	56
17	11	70	49	40	20	65	5	70
18	65	18	85	10	20	58	68	22
19	74	10	95	2	63	13	74	10
20	55	40	42	45	52	23	52	33

We assign the membership functions and non-membership functions to dates A, B, C and D from Table **5.1**. Following conditions must be met, that $\mu, \nu \in \langle 0,1 \rangle$ and $\mu + \nu \leq 1$ for A, B, C and D. Then there are A-IFS data. (Table **5.2**). We calculate the hesitation margins for dates A,B,C a D from following formula $\pi_A(x) = 1 - \mu_A(x) - \nu_A(x)$.

Table 5.2.

A			B			C			D		
μ_A	ν_A	π_A	μ_B	ν_B	π_B	μ_C	ν_C	π_C	μ_D	ν_D	π_D
0.32	0.5	0.18	0.64	0.21	0.15	0.67	0.15	0.18	0.7	0.1	0.2
0.61	0.2	0.19	0.37	0.55	0.08	0.65	0.2	0.15	0.65	0.25	0.1
0.59	0.25	0.16	0.4	0.5	0.1	0.43	0.22	0.35	0.4	0.3	0.3
0.36	0.5	0.14	0.62	0.32	0.06	0.35	0.4	0.25	0.4	0.4	0.2
0.62	0.2	0.18	0.46	0.5	0.04	0.4	0.38	0.22	0.2	0.75	0.05
0.52	0.35	0.13	0.84	0.1	0.06	0.87	0.05	0.08	0.8	0.05	0.15
0.76	0.15	0.09	0.52	0.35	0.13	0.8	0.12	0.08	0.75	0.1	0.15
0.89	0.05	0.06	0.7	0.15	0.15	0.73	0.1	0.17	0.7	0.15	0.15
0.59	0.37	0.04	0.85	0.05	0.1	0.58	0.21	0.21	0.64	0.22	0.14
0.53	0.47	0	0.4	0.54	0.06	0.52	0.37	0.11	0.52	0.37	0.11
0.78	0.12	0.1	0.62	0.37	0.01	0.6	0.18	0.22	0.62	0.28	0.1
0.9	0.05	0.05	0.54	0.38	0.08	0.78	0.05	0.17	0.82	0.1	0.08
0.65	0.24	0.11	0.3	0.51	0.19	0.82	0.02	0.16	0.75	0.13	0.12
0.58	0.4	0.02	0.15	0.65	0.2	0.62	0.15	0.23	0.59	0.24	0.17
0.75	0.12	0.13	0.65	0.17	0.18	0.4	0.45	0.15	0.8	0.12	0.08
0.32	0.6	0.08	0.85	0.08	0.07	0.54	0.38	0.08	0.3	0.56	0.14
0.11	0.7	0.19	0.49	0.4	0.11	0.2	0.65	0.15	0.05	0.7	0.25
0.65	0.18	0.17	0.85	0.1	0.05	0.2	0.58	0.22	0.68	0.22	0.1
0.74	0.1	0.16	0.95	0.02	0.03	0.63	0.13	0.24	0.74	0.1	0.16
0.55	0.4	0.05	0.42	0.45	0.13	0.52	0.23	0.25	0.52	0.33	0.15

Principal Component Analysis

We calculate the correlation matrices for membership R_μ, non-membership values R_ν and hesitation margins R_π (correlation components (5.3) - (5.5) and their eigenvalues and eigenvectors.

$$R_\mu = \begin{pmatrix} 1.00000000 & 0.03399483 & 0.44397730 & 0.6749821 \\ 0.03399483 & 1.00000000 & -0.04322014 & 0.2015552 \\ 0.44397730 & -0.04322014 & 1.00000000 & 0.6576481 \\ 0.67498208 & 0.20155524 & 0.65764810 & 1.0000000 \end{pmatrix}$$

$$R_\nu = \begin{pmatrix} 1.00000000 & 0.08211306 & 0.43986483 & 0.5309616 \\ 0.08211306 & 1.00000000 & -0.06139627 & 0.2920429 \\ 0.43986483 & -0.06139627 & 1.00000000 & 0.6663423 \\ 0.53096160 & 0.29204286 & 0.66634231 & 1.0000000 \end{pmatrix}$$

$$R_\pi = \begin{pmatrix} 1.0000000 & -0.23705667 & 0.17778691 & 0.2057912 \\ -0.2370567 & 1.00000000 & -0.07366807 & 0.1886569 \\ 0.1777869 & -0.07366807 & 1.00000000 & 0.3668935 \\ 0.2057912 & 0.18865693 & 0.36689347 & 1.0000000 \end{pmatrix}$$

The eigenvalues of the correlation matrix R_μ are:

2.2023149, 1.0281328, 0.5496674, 0.2198849.

Variability of input variables (sum of elements on the main diagonal = sum of eigenvalues of the correlation matrix) is $\sigma^2 = 4$. These eigenvalues are illustrated in Fig. (5). From the graph we can see that breaking point is for the second component. Then according to the Kaiser rule the first two components come into consideration.

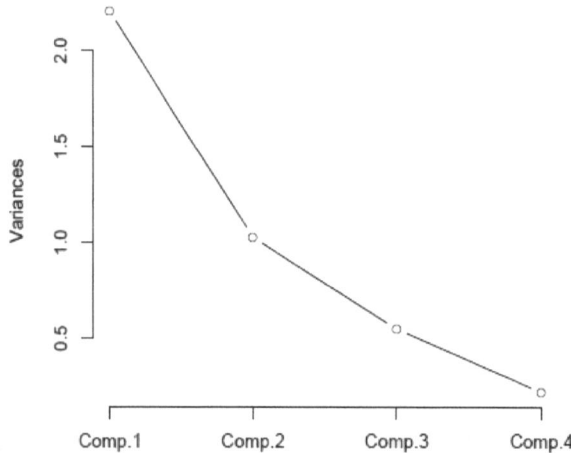

Source: The author of the figure is R. Bartková. The figure has not been published before in any other publication.

Fig. (5). The eigenvalues for correlation matrix R_μ.

We present results PCA obtained through the program R:

Importance of components:

	Comp.1	Comp. 2	Comp. 3	Comp. 4
Standard deviation	1.484019	1.013968	0.741395	0.4689188
Proportion of Variance	0.550578	0.257033	0.137416	0.0549712
Cumulative Proportion	0.550578	0.807611	0.945028	1.0000000

In line "Standard deviation" are values of the variance of principal components $\left(\sqrt{\lambda_i}, i = 1,2,3,4\right)$.

In line "Proportion of Variance" are proportions $\frac{\lambda_i}{\sigma^2}, i = 1,2,3,4$.

In line "Cumulative Proportion" are cumulative proportions of the variability. We can see that three first components explain 80,7% of the overall variation.

Similarly we proceed to non-membership values and hesitation margins.

The eigenvalues for correlation matrix R_μ are :

2.1286980, 1.0511994, 0.5800064, 0.2400963.

Variability of input variables is $\sigma^2 = 4$. The eigenvalues are illustrated in the Fig. (**6**). According to the Kaiser rule we take into consideration the first two components.

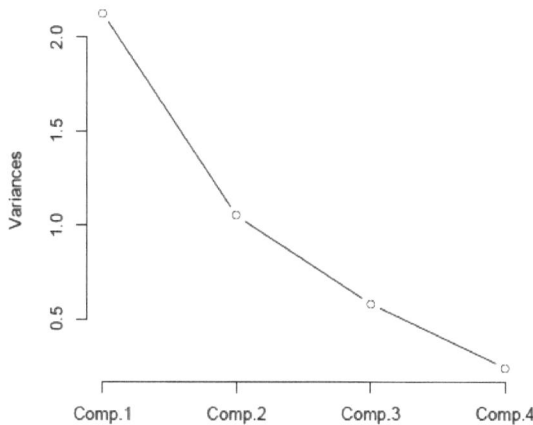

Source: The author of the figure is R. Bartková. The figure has not been published before in any other publication.

Fig. (6). The eigenvalues for correlation matrix R_v.

Importance of components:

	Comp.1	Comp. 2	Comp. 3	Comp. 4
Standard deviation	1.459005	1.025280	0.761581	0.4899961
Proportion of Variance	0.532174	0.262799	0.145001	0.0600240
Cumulative Proportion	0.532174	0.794974	0.939975	1.0000000

We can see that two first components explain 79,5% of the overall variation.

The eigenvalues for correlation matrix R_μ are :

1.5146290, 1.2168545, 0.7610724, 0.5074441.

Variability of input variables is $\sigma^2 = 4$. The eigenvalues are illustrated in Fig. (7). According to the Kaiser rule we take into consideration the first two components. But according to the graph we can see that breaking point is per third component.

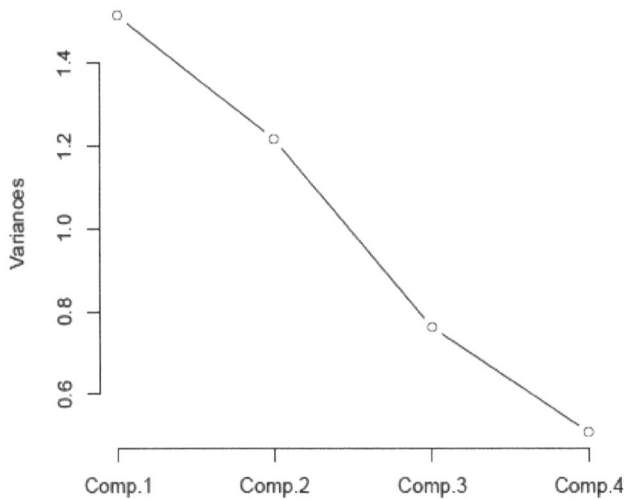

Source: The author of the figure is R. Bartková. The figure has not been published before in any other publication.

Fig. (7). The eigenvalues for correlation matrix R_π.

Importance of components:

	Comp.1	Comp. 2	Comp. 3	Comp. 4
Standard deviation	1.230702	1.103111	0.872394	0.712351
Proportion of Variance	0.378657	0.304213	0.190268	0.126861
Cumulative Proportion	0.378657	0.682870	0.873139	1.000000

We can see that first two components explain 68,29% of the overall variation, which is not sufficient. The first three components explain 87,31% of the overall variation. It is permissible.

Now, we calculate the total correlation matrices R (correlation components (5.2)) and their eigenvalues and eigenvectors.

$$R = \begin{pmatrix} 1.00000000 & -0.04031626 & 0.35387635 & 0.4705783 \\ -0.04031626 & 1.00000000 & -0.05942816 & 0.2274183 \\ 0.35387635 & -0.05942816 & 1.00000000 & 0.5636280 \\ 0.47057831 & 0.22741834 & 0.56362796 & 1.0000000 \end{pmatrix}$$

The eigenvalues for total correlation matrix R are :

1.9385953, 1.0683501, 0.6544004, 0.3386543

The eigenvalues are illustrated in Fig. (**8**).

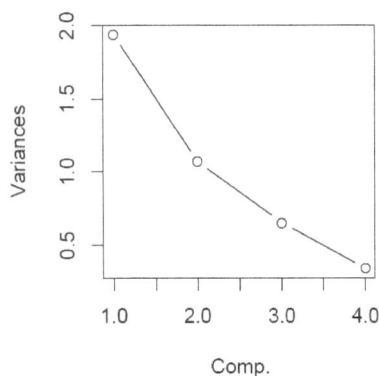

Source: The author of the figure is R. Bartková. The figure has not been published before in any other publication.

Fig. (8). The eigenvalues for the correlation matrix R.

From the graph we can see that breaking point is for the second component. According to the Kaiser rule we take into consideration the first two components. These components explain 75,17% of the overall variation.

Now, we can determine number of principal components which is sufficient for presentation of original four variables. Original four components can be replaced by two principal components while maintaining 75,17% of the variability of original dates.

The result of PCA procedure is in the Table **5.3**. Columns in the table are three first eigenvectors of the covariance matrices R_μ, R_ν, R_π. Principal components is obtained by multiplying the eigenvectors with the original dates.

Table 5.3.

	"membership"		"non-membership"		"hesitation"	
	Principal Components					
	1.	**2.**	**1.**	**2.**	**1.**	**2.**
A	0.55	0.06	-0.52	0.09	0.49	-0.43
B	0.10	-0.97	-0.17	-0.93	-0.09	0.80
C	0.54	0.23	-0.56	0.33	0.61	0.07
D	0.62	-0.09	-0.62	-0.12	0.60	0.41

In this way we obtain a reduced description of the problem in 2 instead of 4 dimensions.

Factor Analysis

We will deal with cases of the factor analysis based on the method PCA. The input data are in Table **5.2**. The correlation matrices and their eigenvalues we have calculated in the previous method PCA. Since that the set of source set have 20 variables, we have at least 2 criteria to determine the number of factors.

The criterion of the eigenvalues determines 2 significant factors (see eigenvalues of correlation matrices R_μ, R_ν, R_π, only the first two are larger than 1, in any case). From graphs of eigenvalues (Figs. **5-7**) we can see that breaking point is per second component.

Further, we examine the overall variation of source set. If we consider two factors then is the overall variation of data equal 75%. Such percentage of variation is sufficient in the case, that we consider data set from the social sciences. Further analysis, we will consider two factors.

At first, we solve the case of FA for data of the membership function $\mu_A, \mu_B, \mu_C, \mu_D$. First two factors represent 80,8% overall variation. Factor analysis performed by using the program R. The matrix of factor loadings is:

$$A_\mu^* = \begin{pmatrix} 0.8213470 & 0.06529165 \\ 0.1495805 & -0.98058653 \\ 0.8045785 & 0.22929373 \\ 0.9262738 & -0.09871300 \end{pmatrix}$$

We can see from the matrix A_μ^*, that the first factor (first column of matrix) have high loadings in the 1st, 3rd and 4th variable. Second factor (second column of matrix) has high loadings in the 2nd variable. We can say that the matrix has simple structure because the matrix has high factor loadings in only one factor. Then the matrix does not necessarily rotate. Values of communalities are:

0.6788739, 0.9839243, 0.6999222, 0.8677274

The values are sufficient. Then we can use two factors instead of original variables.

Next, we solve the case of FA for data of the nonmembership function v_A, v_B, v_C, v_D. First two factors represent 79,5% overall variation. The matrix of factor loadings is:

$$A_v^* = \begin{pmatrix} -0.7622182 & 0.0912679 \\ -0.2450798 & -0.9562097 \\ -0.8178901 & 0.3369507 \\ -0.9048277 & -0.1224619 \end{pmatrix}$$

We can say that matrix A_v^* has simple structure. Values of communalities are:

0.5893063, 0.974401, 0.78248, 0.83371

The values are sufficiently great. Then we can use two factors instead of original variables.

Next, we solve the case of FA for data of hesitation margins $\pi_A, \pi_B, \pi_C, \pi_D$. First two factors represent 68,3% overall variation. The matrix of factor loadings is:

$$A_\pi^* = \begin{pmatrix} 0.6132218 & -0.47283838 \\ -0.1190048 & 0.88433688 \\ 0.7577326 & 0.07817622 \\ 0.7417999 & 0.45289640 \end{pmatrix}$$

We can see from the matrix, that first factor have high loadings in the 3rd and 4th variable. Second factor has high loadings in the 2nd variable. In the first variable the factor loadings is not sufficiently high for any factor. We have to rotate the matrix. The rotate matrix of factor loadings is:

$$A_\pi^{*rot} = \begin{pmatrix} 0.411 & -0.656 \\ 0.195 & 0.871 \\ 0.738 & -0.190 \\ 0.853 & 0.167 \end{pmatrix}$$

In the first variable the factor loadings is not sufficiently high for any factor. We will calculate values of communalities:

0.5996172, 0.7962139, 0.5802703, 0.7553822

The values are sufficiently great. Then we can use two factors instead of original variables.

We remove the first variable from data set and again we calculate FA without this variable.

In this case, first two factors represent 81,5% overall variation. The matrix of factor loadings is:

$$AA_\pi^* = \begin{pmatrix} 0.2729327 & 0.9171304 \\ 0.7609551 & -0.4536824 \\ 0.8560128 & 0.1108828 \end{pmatrix}$$

The rotate matrix of factor loadings is:

$$AA_\pi^{*rot} = \begin{pmatrix} & 0.957 \\ 0.856 & -0.227 \\ 0.793 & 0.342 \end{pmatrix}$$

We can say that the matrix A_v^* has simple structure. A missing value in the matrix is a number close to zero. Values of communalities are:

0.9156203, 0.7848804, 0.7450528

The values are sufficiently great. We have confirmed the theory that we can use two factors instead four original variables.

In this way we obtain a reduced description of the problem in 2 instead of 4 dimensions.

Conclusions

Both methods allow dimensionality reduction of original dataset from 4 to 2 while maintaining the sufficient variability of the original data. In case of PCA method overall variation is 75,17%. In case of FA method overall variation is 81,5%. If we consider datasets from social sciences, then such a percentage of variability is sufficient.

BIBLIOGRAPHY

[1]   Anděl, J. Matematická statistika. Praha : SNTL/ALFA, 1985. ISBN 04-003-85.

[2]   Atanassov, K. Intuitionistic Fuzzy Sets, Springer Physic-Verlag Berlin, 1999.

[3]   Atanassov, K. On Intuitionistic Fuzzy Sets Theory. Springer, Berlin, 2012.

[4]   Atanassov, K.,Tcvetkov, R. On £ukasiewicz's intuitionistic fuzzy disjunction and conjunction. In: Annual of "Informatics" Section, Union of Scientists in Bulgaria, Vol.3, 2010, pp. 90-94.

[5]   Birkhoff, G., von Neumann, J. The logic of quantum mechanics. In: Annals of Mathematics, 37, 1936, pp. 823-834, ISSN 0003-486X

[6]   Bartková, R. Principal Component Analysis and Factor Analysis for IF data sets. In: New Developments in Fuzzy Sets, Intuitionistic Fuzzy sets, Generalized Nets and Related Topics. Volume I: Foundations, IBS PAN - SRI PAS, Warsaw, 2013, pp. 17 - 30.

[7]   Bartholomew, D. J., Steele, F., Galbraith, J., Moustaki, I. Analysis of Multivariate Social Science Data (2 ed.). New York: Chapman & Hall/Crc. 2008.

[8]   Bukowski, P.; Kacprzyk, J.; Szmidt, E. Advances in Principal Component Analysis for Intuitionistic Fuzzy Data Sets, 2012 IEEE 6th International Conference 'Intelligent Szstems' 2012.

[9]   Cushen, C.D., Hudson, R.L. A quantum mechanical central limit theorem. In: Journal of Applied Probability 8, 1971, pp. 454-469, ISSN 00219002.

[10]   Dvurečenskij, A. The Radon-Nikodým theorem for fuzzy probability spaces. In: Fuzzy Sets and Systems. 1992 vol. 45, pp. 69-78, ISSN 0165-0114

[11]   Dvurečenskij, A. On a representation of observables on fuzzy measurable spaces. In: Journal of Mathematical Analysis and Applications, vol. 197, 1996, pp. 579-585, ISSN 0022-247X.

[12]   Dvurečenskij, A. On the existence of probability measures on fuzzy measurable spaces. In: Fuzzy Sets and Systems vol. 43, 1991, pp. 173-181, ISSN 0165-0114.

[13]   Dvurečenskij, A. Metódy teórie F-kvantových priestorov (Methods of the theory of F-quantum spaces). Teória a aplikácie fuzzy množín IV-VII, VVTŠ Liptovský Mikuláš, 1989, pp. 27-37.

[14]   Dvurečenskij, A. Signed states on a logic. In: Mathematica Slovaca 28, 1978, pp. 33-40, ISSN 0139 - 9918.

[15]   Dvurečenskij, A. On a some properties of transformations of a logic. In: Mathematica Slovaca 26, No. 2, 1976, pp. 131-137, ISSN 0139 - 9918.

[16]   Dvurečenskij, A., Chovanec, F. Compatibility and summability of observables in fuzzy quantum spaces. In: BUSEFAL 1993, vol. 56, pp. 4-14, ISSN 0296-3698.

[17]   Dvurečenskij, A., Chovanec, F. Fuzzy quantum spaces and compatibility. In: International Journal of Theoretical Physics 1988, vol. 27, pp. 1069-1082, ISSN 0020-7748.

[18]   Dvurečenskij, A., Chovanec, F., Kôpka, F. On mean value additivity on fuzzy quantum spaces. In: Acta Mathematicae Univ. Comenianae 1990, vol. 58-59, pp. 107-117, ISSN 0862-9544.

[19]   Dvurečenskij, A., Kôpka, F., Riečan, B. On a representation theorem of observables in ordered spaces. BUSEFAL, 56, 1993, pp. 15-19, ISSN 0296-3698.

[20]   Dvurečenskij, A., Pulmanová, S. On joint distribution of observables. In: Mathematica Slovaca, 32, 1982, pp. 155-166, ISSN 0139 - 9918.

[21] Dvurečenskij, A., Riečan, B. On joint distribution of observables for F-quantum spaces. In: Fuzzy Sets and Systems 20, No. 1, 1991. 65-73, ISSN 0165-0114.

[22] Dvurečenskij, A., Riečan, B.: Fuzzy quantum models. In: Int. J. of General Systems 20, 1991, No. 1, pp. 39-54, ISSN 0308-1079.

[23] Dvurečenskij, A., Riečan, B. On joint observables for F-quantum spaces. In: BUSEFAL 35, 1988, pp. 10-14, ISSN 0296-3698.

[24] Embrechts, P.; Kluppelberg, C., Mikosch, T. Modelling Extremal Events: For Insurance and Finance. Springer-Verlag. 1997, ISSN 0172-4568, p. 152-180.

[25] Grzegorzewski, P.,Mrówka, E. Probability on intuitionistic fuzzy sets. In: Soft Methods in Probability, Statistics and Data Analysis. Advances in Intelligent and Soft Computing Volume 16, Springer, 2002, pp. 105-115.

[26] Gudder, S. Joint distributions of observables. In: Journal of Mathematics and Mechanics 18, 1968, pp. 325-335, ISSN 0095-9057.

[27] Gudder, S., Mullikin, H.C. Measure theoretic convergences of observables and operators. Journal of Mathematical Physics 14, 1973, pp. 234-242, ISSN 0022-2488.

[28] Gumbel, E. J. Statistics of Extremes. New York: Columbia University Press, 1958, ISBN 0-486-43604-7.

[29] Gunson, J. On the algebraic structure of quantum mechanics. Communications in Mathematical Physics 6, 1967, pp. 262-285, ISSN 0010-3616.

[30] Guz, W. Fuzzy -algebras of physics. Int. Journal of Theoretical Physics, 24, 1985, pp. 481-493, ISSN 0020-7748.

[31] Haan, L., Ferreira, A. Extreme Value Theory: An Introduction. Springer, 2006.

[32] Halmos, P.R. Lectures on Ergodic Theory. New York: Chelsea Pub. Co. AMS ISBN 978-0-8218-4125-9

[33] Halmos, P.R. Measure theory. Van Nostrand, New York 1958.

[34] Harman, B., Riečan, B. On the individual ergodic theorem in F-quantum spaces. In: Zeszyty naukove Akademii Ekonomicznej w Poznaniu - seria 1, 187, 1992, pp. 25-30.

[35] Heisenberg, W. Über quantentheoretische Umdeutung kinematischer und mechanischer Beziehungen. Zeitschrift für Physik 33, 1925, pp. 879-893, ISSN 0939-7922.

[36] Chovanec, F., Kôpka, F. Fuzzy equality and convergences for F-observables in F-quantum spaces. In: Applications of mathematics 36, 1991, No. 1, pp. 32-45, ISSN 0802-794.

[37] Jajte, R. Strong laws of large numbers for several contractions in a von Neumann algebra. In: Lecture Notes in Mathematics. Vol. 1391, 1989, pp. 125-139, ISSN 0075-8434

[38] Kacprzyk, J., Szmidt, E. Correlation of Intuitionistic Fuzzy Sets, Lecture Notes in AI, 6178, 2010, pp. 169-177.

[39] Kolmogorov, A.N. Osnovnyje ponjatija teorii verojatnostej. Moskva: Nauka 1974, 119 pp.

[40] Kôpka, F., Chovanec, F.: Martingale convergence theorem in F-quantum spaces. In: BUSEFAL, 56, 1993, pp. 20-28, ISSN 0296-3698.

[41] Lendelová, K. Conditional IF-probability. Soft Methods for Integrated Uncertainty Modelling. Volume 37 of the series Advances in Soft Computing 2006, pp. 275-283, ISBN 978-3-540-34776-7 (Print) 978-3-540-34777-4 (Online)

[42] Mac Laren, M. Atomic orthocomplemented lattices. In: Pacific Journal of Mathematics 14, 1964, pp. 597-612, ISSN 0030-8730.

[43]   Mackey, G. W. Mathematical Foundations of Quantum Mechanics. New York - Amsterdam 1963, ISBN 0-486-43517-2.

[44]   Markechová, D. Fuzzy Probability Spaces, Fuzzy dynamical Systems, and Entropy. LAP LAMBERT Academic Publishing, Germany, 2013, 107 p., ISBN 978-3-659-46334-1.

[45]   Markechová, D. F-dynamické systémy a ich entrópia (F-dynamical systems and their entropy). FPV UKF v Nitre, Edícia: Prírodovedec n.447, 2011, 80 p., ISBN 978-80-8094-884-9.

[46]   Markechová, D. The entropy on F-quantum spaces. In. Mathematica Slovaca, Vol 40, 1993, pp. 177-190, ISSN 0139-9918.

[47]   Markechová, D. The entropy of fuzzy dynamical systems and generators. In: Fuzzy Sets and Systems, Vol. 48, no. 3, 1992, pp. 351-363, ISSN 0165-0114.

[48]   Markechová, D. Isomorphism and conjugation of fuzzy dynamical systems. In: BUSEFAL, 38, 1989, pp. 94-101, ISSN 0296-3698.

[49]   Markechová, D. The entropy of fuzzy dynamical systems. In: BUSEFAL, 38, 1989, pp. 38-41, ISSN 0296-3698.

[50]   Markechová, D., Tirpáková, A. The Lebegue decomposition theorem for fuzzy quantum spaces. In: Fuzzy Sets and Systems, 1994, pp. 203-210, ISSN 0165-0114.

[51]   Markechová, D., Tirpáková, A. Birkhoff's individual ergodic theorem and maximal ergodic theorem for fuzzy dynamical systems, 2016. DOI 10.1186/s13662-016-0855-x. In: In. Advances in Difference Equations, 2016, online, p. 1-8, ISSN 1687-1847 <https://advancesindifferenceequations.springeropen.com/articles/   10.1186/s13662-016-0855-x>

[52]   Mrakechová, D., Riečan, B. Entropy of Fuzzy Partitions and Entropy of Fuzzy Dynamical Systems. In: Entropy, 2016, vol 18 (1),19 doi 10.3390/e18010019, ISSN 1099-4300, < http://www.mdpi.com/1099-4300/18/1/19>

[53]   Mesiar R. Fuzzy probability measures and the Bayes formula. In Electronic BUSEFAL 56, 1993.

[54]   Mesiar R.Symmetric fuzzy probability measures and the Bayes formula. In: BUSEFAL 34, 1988, pp. 130-137, ISSN 0296-3698.

[55]   Mesiar R. Fuzzy Bayes formula and the ill defined elements. In: Proceeding from the First Winter School on Measure Theory, Lipt. Ján, 1988, pp. 82-87.

[56]   Mesiar R. A generalization of the individual ergodic theorem. In: Mathematica Slovaca 30, 1980, pp. 327-330, ISSN 0139-9918.

[57]   Mittelstaedt, P. Sprache und Realität in der modern Physic. BI-Wissenschaftsverlag, Manheim, 1986, 258 p.

[58]   Mundici, D. Interpretation of AFC*-algebras in Lukasiewicz sentential calculus. In: J. Funct. Anal., 1986, vol. 6, p. 889-894.

[59]   Navara, M. Existence of states on quantum structures. In: Information Sciences 179, No. 5, 2009, pp. 508-514, ISSN 0020-0255.

[60]   Navara, M., Pták, P. P-measures on soft fuzzy σ-algebras. In: Fuzzy Sets and Systems 56, No. 1, 1993, pp. 123-126, ISSN: 0165-0114.

[61]   Navara, M., Pták, P. On the state space of soft fuzzy algebras. In: BUSEFAL 48, 1991, pp. 55-63, ISSN 0296-3698.

[62]   Neubrunn, T., Riečan, B. Miera a integrál (Measure and integral). Veda, Bratislava, 1981, 555 p.

[63]   Von Neumann, J. Mathematische Grundlagen der Quantenmechanik. Springer-Verlag Berlin - Heidelberg - New York, 1932, 1968, 1996) Printed in Germany, 255 p., ISBN 3-540-59207-5.

[64]   Ochs, W. On strong law of large numbers in quantum probability theory. In: Journal of Philosophical Logic, Vol. 6, 1977, pp. 473-480, ISSN 0022-3611.

[65]   Ochs, W. Concepts of convergence for a quantum law of large numbers. In: Reports on Mathematical Physics, vol. 17 (1980), 127-143, ISSN 0034-4877.

[66]   Palumbíny, O. Zákony ve¾kých èísel a centrálne limitné vety na F-kvantových priestoroch (Law of large numbers and central limit theorems on F-quantum spaces). Thesis, Comenius University, 1988.

[67]   Piasecki, K. On fuzzy P-measure. In: Proc. First Winter School Measure Theory, Liptovský Ján, 1988, pp. 108-112.

[68]   Piasecki, K. Note on „On the Bayes formula for fuzzy probability measure". In: Fuzzy Sets and Systems 24, 1987, pp. 121-122, ISSN 0165-0114.

[69]   Piasecki, K. On the Bayes formula for fuzzy probability measure. In: Fuzzy Sets and Systems 18, 1986, pp. 183-185, ISSN 0165-0114.

[70]   Piasecki, K. Probability of fuzzy events defined as denumerable additivity measure, In: Fuzzy Sets and Systems 17, 1985, pp. 271–284, ISSN 0165-0114.

[71]   Piasecki, L., Svitalski, Z. A remark on the definition of fuzzy P-measure and the Bayes formula. In: Fuzzy Sets and Systems 26, 1988, pp. 112-124, ISSN 0165-0114.

[72]   Pulmanová, S. Individual ergodic theorem on logic. In: Mathematica Slovaca 32, 1982, pp. 413-415, ISSN 0139-9918.

[73]   Pykacz, J. Quantum logics and soft fuzzy probability spaces. In: BUSEFAL 32, 1987, pp. 150-157, ISSN: 0296-3698.

[74]   Renčová, M. On the E-Probability on IF-Events. In. Lecture Notes in Computer Science. Volume 5571 2009, "Fuzzy Logic and Applications" 8th International Workshop, WILF 2009 Palermo, Italy, June 9-12, 2009 ISBN: 978-3-642-02281-4 (Print) 978-3-642-02282-1, <http://link.springer.com/book/10.1007%2F978-3-642-02282-1>

[75]   Renčová, M. A generalization of probability theory on MV-algebras to IF-events. In: Fuzzy Sets and Systems, Volume 161, Issue 12, 16 June 2010, pp. 1726-1739, ISSN 0165-0114

[76]   Révész, P. The laws of large numbers. Academic Press Inc, 1968, ISBN-10: 0125871503

[77]   Riečan, B. On finitely additive IF-states. In: Intelligent Systems´2014, Springer, Berlin, 2014, pp. 149 – 15, ISBN 978-3-319-11312-8.

[78]   Riečan, B. A descriptive definition of the probability on intiutionistic fuzzy sets. In: EUSFLAT'2003 (M.Wagenecht, R.Hammpel, eds.), Univ. Appl. Sci., Zittau-Goerlitz 2003, p. 263 - 266.

[79]   Riečan, B. A new approach to some notions of statistical quantum mechanics, Busefal, 1988, 35, p. 4-6, ISSN 0296-3698.

[80]   Riečan, B. On the convergence of observables in fuzzy quantum logics. In: Tatra Moutains Mathematical Publications 6, 1995, pp. 149-156, ISSN 1210 – 3195.

[81]   Riečan, B. On some mathematical models of quantum mechanical systems. BUSEFAL 56, 1993, pp. 63-66, ISSN: 0296-3698.

[82] Riečan, B. On mean value in F-quantum spaces. In: Aplikace matematiky 35, 1990, No. 3, pp. 209-214, ISSN 0862-7940.

[83] Riečan, B. Indefinite integral in fuzzy quantum spaces. BUSEFAL 38, 1989, 5-7, ISSN 0296-3698.

[84] Riečan, B. O pravdepodobnosti na fuzzy množinách (About the probability on fuzzy sets). In: Proc. Probastat'89, Liptovský Ján, 1989, 35-94.

[85] Riečan, B. Strong Poincaré recurrence theorem in MV-algebras. In: Mathematica Slovaca, Volume 60, Issue 5 (Oct 2010), pp.. 655-664, ISSN 1337-2211

[86] Riečan, B. On local representation of some algebraic structures. In. Proc. 4th International IEEE Conference Intelligent Systems, 09/2008, 13/21 - 13/23

[87] Riečan, B. On the Probability Theory on the Atanassov Sets. Chapter in Intelligent Techniques and Tools for Novel System Architectures, Volume 109 of the series Studies in Computational Intelligence, pp. 395-413, ISBN 978-3-540-77621-5 (Print) 978-3-540-77623-9 (Online)

[88] Riečan, B. On the Atanassov Concept of Fuzziness and One of Its Modification. In: Imprecision and Uncertainty in Information Representation and Processing. Studies in Fuzziness and Soft Computing, New Tools Based on Intuitionistic Fuzzy Sets and Generalized Nets, Volume 332, 2016, pp. 27-40, ISBN 978-3-319-26301-4 (Print) 978-3-319-26302-1 (Online)

[89] Riečan, B. Probability theory and the operations with IF-sets. IEEE International Conference on Fuzzy Systems (IEEE World Congress on Computational Intelligence), 2008, pp. 1250 - 1252, DOI: 10.1109/FUZZY.2008.4630531 http://ieeexplore.ieee.org/search/searchresult. jsp?searchWithin=%22 Authors%22:.QT.Beloslav%20Riecan.QT.&newsearch=true

[90] Riečan, B., Atanassov K. Some properties of operations conjunction and disjunction from Lukasiewicz type on intuitionistic fuzzy sets. Part 1. Notes on Intuitionistic Fuzzy Sets 20, 2014, No. 3, pp. 1-6, ISSN 1310-4926.

[91] Riečan, B., Ciungu, L. General Form of Probabilities on IF-Sets. Fuzzy Logic and Applications, 2009, Volume 5571 of the series Lecture Notes in Computer Science pp. 101-10,. 8th International Workshop, WILF 2009 Palermo, Italy, June 9-12, 2009 Proceedings, ISBN 978-3-642-02281-4 (Print) 978-3-642-02282-1 (Online) <http://link.springer.com/ book/10.1007/978-3-642-02282-1>

[92] Riečan, B., Neubrunn, T. Teória miery (Measure theory). Veda, Bratislava, 1992, ISBN 978-80-224-0368-9

[93] Riečan, B., Neubrunn, T. Integral, Measure and Ordering. Kluwer. Dordrecht, 1997, ISBN 80-88683-18-1.

[94] Rüttimann, G.T. Hahn-Jordan decomposition of signed weight on Finite orthogonality spaces. In: Commentarii mathematici Helvetici 52, 1977, pp. 129-144, ISSN 0010-2571.

[95] Samuelčík, K. Conditional probability on the Kôpka's D-posets. In: Acta Mathematica Sinica, English Series, November 2012, Volume 28, Issue 11, pp. 2197-2204, ISSN 1439-8516 (Print) 1439-7617 (Online)

[96] Schrodinger, E. Quantisierung als Eigenwertproblem. In: Annalen der Physik, Vol. 384, 1926, pp. 361-376, Online ISSN 1521-3889.

[97] Sikorski, R. Boolean algebras. Berlin, Springer-Verlag, 1964.

[98] Szmith, E. Distances and Similarities in Intuitionistic Fuzzy Sets. Springer International publishing Seitzerland 2014, ISBN:978-3-319-01639-9 (Print) 978-3-319-01640-5 (Online)

[99]   Šerstnev, A.N.: O poňatii zarjada v nekomutativnej scheme teorii mery. Veroj. Met. I kiber. N. 10-11, Kazaň, 1974, pp. 66-77.

[100] Tirpáková, A. 1990. *Príspevok k teórii pravdepodobnosti na F-kvantových priestoroch (A contribution to the theory of probability on F-quantum spaces)* Dissertation thesis. Bratislava: MFF KU, 1990.

[101] Tirpáková, A., Markechová, D. Signed measures and Hahn–Jordan decompositions of fuzzy measurable spaces. In: Demonstratio Mathematica, Vol. 29, no. 1, 1996, pp. 37-42, ISSN 0420-1213.

[102] Tirpáková, A., Markechová, D. The fuzzy analogies of some ergodic theorems 2015. DOI 10.1186/s13662-015-0499-2. In: Advances in Difference Equations, Vol. 2015, no. 1 (2015), article number 171, p. 1-10, ISSN 1687-1839 <https://advancesindifferenceequations. springeropen.com/articles/ 10.1186/s13662-015-0499-2>

[103] Varadarajan, V.S. Geometry of quantum theory. Second edition, Springer, 2007.

[104] Varadarajan, V.S. Probability in physics and a theorem on simultaneous observability. In: Communications on Pure and Applied Mathematics 15, 1962, pp. 189-217, ISSN 0010-3640.

[105] Zadeh, L.A. Fuzzy sets. In: Information and Control 8, No. 3, 1965, pp. 338-353, ISSN 0890-5401.

[106] Zadeh, L.A. Probability measures of fuzzy events. In: Journal of Mathematical Analysis and Applications, Vol. 23, No. 2, 1968, pp. 421-427, ISSN 0022-247X.

[107] Zadeh, L.A. Fuzzy Sets, Fuzzy Logic, and Fuzzy Systems: Selected Papers by Lotfi A. Zadeh. World Scientific Pub Co Inc; First Edition edition, 1996, ISBN-13: 978-9810224219.

SUBJECT INDEX

W

Z

www.ingramcontent.com/pod-product-compliance
Lightning Source LLC
Chambersburg PA
CBHW041727210326
41598CB00008B/805